Ancestral Places

Other volumes in the **First Peoples: New Directions in Indigenous Studies** series

Accomplishing NAGPRA: Perspectives on the Intent, Impact, and Future of the Native American Graves Protection and Repatriation Act
Edited by Sangita Chari and Jaime M. N. Lavallee

Asserting Native Resilience: Pacific Rim Indigenous Nations Face the Climate Crisis
Edited by Zoltán Grossman and Alan Parker

A Deeper Sense of Place: Stories and Journeys of Collaboration in Indigenous Research
Edited by Jay T. Johnson and Soren C. Larsen

The Indian School on Magnolia Avenue: Voices and Images from Sherman Institute
Edited by Clifford E. Trafzer, Matthew Sakiestewa Gilbert, and Lorene Sisquoc

Salmon Is Everything: Community-Based Theatre in the Klamath Watershed
Theresa May with Suzanne Burcell, Kathleen McCovey, and Jean O'Hara

Songs of Power and Prayer in the Columbia Plateau: The Jesuit, The Medicine Man, and the Indian Hymn Singer
Chad S. Hamill

To Win the Indian Heart: Music at Chemawa Indian School
Melissa D. Parkhurst

Ancestral Places

Understanding Kanaka Geographies

Katrina-Ann R. Kapāʻanaokalāokeola Nākoa Oliveira

FIRST PEOPLES
New Directions in Indigenous Studies

Oregon State University Press *Corvallis*

The paper in this book meets the guidelines for permanence and durability of the Committee on Production Guidelines for Book Longevity of the Council on Library Resources and the minimum requirements of the American National Standard for Permanence of Paper for Printed Library Materials Z39.48-1984.

Library of Congress Cataloging-in-Publication Data
Kapāʻanaokalāokeola Oliveira, Katrina-Ann Rose-Marie.
Ancestral places : understanding Kanaka geographies / Katrina-Ann R. Kapāʻanaokalāokeola Nākoa Oliveira.
pages cm. — (First peoples : new directions in indigenous studies)
Includes bibliographical references and index.
ISBN 978-0-87071-673-7 (alk. paper) — ISBN 978-0-87071-674-4 (e-book)
1. Hawaiians—Ethnic identity. 2. Hawaiians—Psychology. 3. Hawaiians—Social life and custom. 4. Names, Geographical—Hawaii. 5. Hawaiian language—Name. 6. Geographical perception—Hawaii. 7. Place attachment—Hawaii. 8. Environmental geography—Hawaii. 9. Hawaii—Geography. I. Title.
DU624.65.K38 2014
919.69—dc23
2013041310

 First published in 2014 by Oregon State University Press
Printed in the United States of America

Oregon State University Press
121 The Valley Library
Corvallis OR 97331-4501
541-737-3166 • fax 541-737-3170
www.osupress.oregonstate.edu

Dedication

No kuʻu mau kūpuna a me kuʻu mau mākua i hehi i ke alanui kīkeʻekeʻe ma mua oʻu, no kuʻu kaikaina lāua ʻo kuʻu kaikunāne e hele kūnihi like aku nei i nēia ala ma ka ʻaoʻao oʻu, a no kuʻu kama nō hoʻi e neʻe aku ana i mua ma hope oʻu.

No kuʻu ēwe, kuʻu piko, kuʻu iwi, kuʻu koko. E ola nā iwi, e ola nā kulāiwi, e ola nā koa, e kū kahiko!

Contents

Acknowledgments

It is with a grateful and humble na'au that I thank my kūpuna for the wealth of knowledge and legacy of excellence with which they gifted succeeding generations. I thank my mother, Maryann Nākoa Barros, who has been my strongest supporter on this journey, giving selflessly of her time and energy so that I could complete this book. I thank my fathers, Frank G. Oliveira and Jacob K. Barros, who both passed away much too early, for all of their love, support, and wisdom. To my brother Jared Barros, my sister Joylynn Paman, my brother-in-law Daniel Paman, my nephews Ka'imipono and Kamaha'o Paman, and my niece 'Ilikea Paman, I thank you for keeping me grounded. To my kāne, Alika Apo, thank you for supporting me through this stressful time in our lives. To my daughter, Kahakai Apo, my most precious blessing, I thank you for your unconditional love and for your endless smiles, kisses, and hugs. To my many aunties, uncles, cousins, and friends, who stood there by my side throughout this process, mahalo a nui loa for your love and support. I would especially like to thank my Barros, Nākoa, and Oliveira 'ohana.

To my mentors on this project, Brian Murton and Noenoe Silva, I extend to you my deepest gratitude for keeping me on track and assisting me to see the light at the end of the tunnel. Without your guidance, this project would never have come to fruition. A special mahalo also goes out to Lilikalā Kame'eleihiwa, Mary McDonald, and Everett Wingert for serving on my doctoral committee and for assisting me with the framework that would eventually lead to the publication of this book.

To Mary Elizabeth Braun, acquisitions editor for Oregon State University Press, thank you for believing in me and assisting me through this arduous process. To Jo Alexander and Micki Reaman, managing editors for Oregon State University Press, I extend my heartfelt gratitude to you for giving me the time I needed to produce the best book possible. To Lalepa Koga, Kamaoli Kuwada, Lee S. Motteler, Kawailana Saffery, Kekeha Solis, and Laiana Wong, thank you for painstakingly editing my manuscript. To Manu Meyer and the other reviewers of my manuscript, mahalo a nui loa for your constructive and insightful feedback.

Mahalo a nui loa to Haley Kailiehu and Lilinoe Andrews for being lights for me as I traveled down this alanui kīkeʻekeʻe. Haley, I am honored that you allowed me to grace the pages of this book with your awesome illustrations, illuminating key concepts in the text. Lilinoe, I am forever grateful that you came to my aid to format my genealogy charts. You are so talented.

Words cannot express how indebted I am to my colleagues (many of whom are my former professors), Hōkūlani ʻAikau, Carlos Andrade, Maile Andrade, Noelani Arista, Leilani Basham, Kahele Dukelow, Noelani Goodyear-Kaʻōpua, kuʻualoha hoʻomanawanui, Kamuela KaʻAhanui, Kaleikoa Kaʻeo, Manu Kaʻiama, Lalepa Koga, Charles Lawrence III, Renee Louis, Kawehi Lucas, Margie Maaka, Jon Osorio, Alicia Perez, Kawailana Saffery, Kekeha Solis, Kūkini Suwa, Noʻeau Warner, Kalani Wise, Kainoa Wong, Laiana Wong, Kahunawai Wright, and Kanalu Young, for engaging in enlightening discussions and thought-provoking writing group sessions. Without your aloha and support, I would not have had the confidence to even begin to think about publishing this book.

To the staff of the Bishop Museum Archives, I am truly indebted to all of you for the grace and humility with which you serve the community. To Aunty Pat Bacon, who is the source of much ʻike, my life has been enriched by our many conversations in the archives. To the Bishop Museum staff, ʻAnoʻilani Aga, Nanea Armstrong-Wassel, DeSoto Brown, Leah Caldeira, Deanne DuPont, Betty Kam, Judith Kearny, Albert Roland, Ron Schaeffer, and B. J. Short, thank you very much for sharing your time and expertise with me.

Mahalo to Dean Maenette Benham of Hawaiʻinuiākea School of Hawaiian Knowledge for approving my sabbatical and giving me the priceless gift of time.

To the Mellon-Hawaiʻi fellowship and the Kohala Center, thank you for giving me the time and resources to complete this manuscript. A special mahalo goes out to Matt Hamabata and Cortney Okamura for all of your hard work behind the scenes to ensure the success of fellows like me. Mahalo to all of the Mellon-Hawai'i fellows for an awesome network of support, especially my cohort, Alohalani Brown and Kaipo Perez.

Mahalo to the Kamehameha Schools' First Nations' Futures Program for allowing me to spend a year with all of you to discuss important 'āina-based management issues. To B. J. Awa, Kamana Beamer, Ka'eo Duarte, Neil Hannahs, Māhealani Matsuzaki, Kanoe Wilson, and 'Ulalia Woodside, mahalo a nui loa for perpetuating the legacy of our ali'i through your important work. To my Papa Kalū'ulu cohort, Kapi'olani Adams, Pua Fernandez, Keoni Lee, Ka'iulani Murphy, and Lori Tango, I have the deepest respect for all of you and the work that you do for the lāhui. Thank you for your support during this writing process.

Finally, I would like to thank some kupa o ka 'āina, Uncle Sammy Chang, Uncle Oliver Dukelow, Aunty Pi'imauna Dukelow, Uncle Les Dunn, Uncle Eddie Kaanaana, Uncle Harry Mitchell Jr., and Grandpa Ned Kalawai'anui Nākoa Sr., for always reminding me to come home to continue the legacy of my kūpuna.

Introduction

My daughter, Kahakai, was born at the Maui Memorial Hospital in the kau (season) of Hoʻoilo in the mahina (month) of Kaulua on the pō mahina (night) of Akua. I would have preferred to have given birth to her on our kulāiwi (ancestral homeland) on the mokupuni (island) of Maui in the moku (district) of Kāʻanapali in the kalana (smaller division of land than moku) of Kahakuloa in the ʻili (small land division) of Kuewa,[1] but the remoteness of Kuewa might have placed my daughter's life in jeopardy had there been complications during her delivery.

Kahakai's birth was met by a storm that loomed over Maui the days before, during, and after her birth. In fact, Kahakuloa experienced the most severe flooding in the valley in recent memory. It was so severe that the stream breached its banks and flooded places that lifelong residents of Kahakuloa had never seen flooded before—even during hurricanes. Nine days after her birth, my kāne (male partner) and I took Kahakai to Kuewa to kanu (bury, plant) her ʻiewe (placenta) and piko (umbilical cord), just as my ancestors had done for many generations.[2]

It was important to me to give birth to my daughter on the island of Maui, as my kāne and I both have ancestral roots on Maui, our families having resided on Maui since time immemorial. He was born and raised on Maui. I was raised on Maui, where my mother lived, and also on Oʻahu, where my father lived following the divorce of my parents when I was four. As a youth, flying back and forth between both islands every other weekend and spending my summers on Maui was a normal part of my life.

When I started taking ʻōlelo Hawaiʻi (Hawaiian language) classes, I suddenly had an identity crisis. I had to reflect upon who I was and where I came from. When meeting someone for the first time, two of the most common questions asked in ʻōlelo Hawaiʻi deal with place and identity. Often new acquaintances will ask, "No hea mai ʻoe?" (Where are you from?) and "Na wai ke kama ʻo ʻoe?" (Who do you descend from?). While I grew up knowing the names of my mākua (parents) and my kūpuna (ancestors), answering the question "No hea mai ʻoe?" was more complicated. As my kumu ʻōlelo Hawaiʻi (Hawaiian language teacher) explained, a "proper" response to the question is to state the (one) place you were raised.

For many years, the question "No hea mai ʻoe?" proved difficult to answer. In spite of being born on Oʻahu, going to school on Oʻahu, and working on Oʻahu, I felt a much deeper connection to Maui. Growing up on Oʻahu, I lived an urban life. I did not have a deep relationship with the ʻāina (land, that which feeds). I lived in a house on a property that my parents purchased after I was born. I went to school and came home. I did not know my neighbors.

Growing up on Maui, I lived a rural life. I developed a deep appreciation and love for the land. I swam in the streams, worked in the loʻi (wetland taro gardens), fished in the ocean, ate the fruits of the land, and cared for the burial grounds of my kūpuna. I grew up on the kulāiwi of my kūpuna; everyone in the village was either family or family friends.

Since the answer to "No hea mai ʻoe?" is often a singular place, my kāne and I decided that we wanted our daughter to be born on Maui so that she could state, without hesitation, "Ua hānau ʻia au ma Maui. No Maui mai au" (I was born on Maui. I am from Maui). Because I have a job that I am passionate about on Oʻahu, like me, Kahakai will likely reside on two islands throughout her life, but her ʻiewe and her piko are buried on Maui.

Unlike my experience in contempory times, makaʻāinana (the general populace who lived off the land) often lived in the same place throughout their entire lives. As the economy in ka pae ʻāina Hawaiʻi (Hawaiian archipelago) shifted from a barter system to a monetary commerce system, Kānaka (Native Hawaiians)[3] began adapting to a new economy. As a result, over time, many people left their homelands and subsistence farming and fishing lifestyles in search of jobs. The shift in the economic system coupled with the introduction of transportation in the form of horses and automobiles created a more mobile society. Today, I live a highly mobile lifestyle, flying to Oʻahu during the week to work at the university and flying to Maui on weekends to be with my family and my kulāiwi on Maui. Oʻahu is where I was formally trained as an academic; Maui is where I learned many of my life's lessons. I am from both places, but my kuleana differs on each island.

Throughout the course of writing this book, the driving forces behind my writing have been my kūpuna, my mākua, my kama (child), and my ancestral homelands. My kūpuna have inspired me to look to the wisdom of the past, my mākua have challenged me to make the most of the present, my daughter has reminded me to create a legacy for the future, and my ancestral places have grounded me and molded my identity.

Fortunately, some of our ancestral homelands are still cared for by our ʻohana (family). This allowed me the opportunity to write the initial draft of this book in Kuewa. Making the decision to write there was easy; I knew that the only way I could write about Kanaka (Native Hawaiian) geographies, connections to ʻāina, and ancestral knowledge systems was to be on at least one of my kulāiwi. Although deciding to write somewhere without electricity, running drinking water, and other modern conveniences was initially challenging, in the end it proved to be a blessing. Before I could begin the process of writing, however, I needed to mālama ʻāina (care for the land). With the assistance of my family, we cleared the ʻāina so I could drive to the house my grandfather built for the family. I purchased a generator to run my computer, and my parents ran the last available telephone service in the valley to our home. My only housemate was my dog.

Living in Kuewa was a life-changing experience. I became more in touch with my environment, and I began to sense the presence of my kūpuna. While living in Kuewa, I was more dependent on my environment than any other time in my life. If the stream flooded, I could not leave the valley because I could not drive safely through the two stream crossings. If there was a landslide blocking the road, I could not drive to town. If a boulder or wild pigs broke the water line to the house, I did not have piped stream water to wash dishes or flush the toilet.

The experience helped me to gain a deeper respect for the ancestral knowledge systems that my kūpuna developed as a result of their observations and interactions with their environment. Without the experience of living in Kahakuloa Valley, this book could not have been written. I needed to distance myself from outside distractions in order to focus on the book. And, more importantly, I needed to listen to the lessons that the ʻāina and my kūpuna were about to reveal to me.

Taking Wahi Pana o Oʻahu, a Hawaiian Studies course at the University of Hawaiʻi at Mānoa from Kalani Wise was another pivotal point in my life that put me on this alanui kīkeʻekeʻe (zigzag road) and this ala hele kūnihi (precarious path). The late Kalani Wise, an amazing and mesmerizing kumu (teacher, source of knowledge) of Kanaka geography, captivated my attention each day with moʻolelo (historical accounts) related to the island of Oʻahu. For fun, he drove around the island to learn more about Oʻahu's street names and geographical features. With each course lecture, he took me on an adventure around Oʻahu where I learned about land boundaries, place names, and other geographical knowledge that he had researched in the libraries and archives as well as on the landscape itself.

Kalani Wise lived up to his name; he was indeed a very wise man who piqued my interest in geography. Soon after his course was over, I graduated with dual degrees in Hawaiian Studies and Hawaiian Language, and Kalani Wise passed away. After graduating with my Bachelor of Arts degrees, I entered the Master of Arts program in Geography at the University of Hawaiʻi at Mānoa because I was eager to learn more about the Kanaka geographies to which I had been introduced by my kumu, Kalani Wise.

I feel a deep and profound responsibility to my kumu and to my kūpuna to give back to academia and my Kanaka community by sharing whatever little ʻike (knowledge) I have about Kanaka place making. This book is my contribution to the next generation, as I feel an incredible sense of kuleana to carry on the knowledge of my kūpuna. As I am a kumu ʻōlelo Hawaiʻi by profession and a geographer by training, I am blending both disciplines in this book by elucidating a great deal of ʻōlelo knowledge about place making, especially through the use of ʻōlelo noʻeau (wise sayings). It is my hope that this book will make significant contributions to our collective project of regenerating our ʻōlelo and the practices of our kūpuna by demonstrating to the next generation that our native tongue and our ancestral knowledge systems are still relevant to us today.

This book acknowledges the wisdom of our kūpuna and demonstrates the ways in which their wisdom continues to inform our identity as Kānaka. My aim is to be original in my interpretations of Kanaka geographies, yet beholden to those who have taken this path previously traveled by my kūpuna. It is hoped that this book will encourage other Kānaka to (re)connect to their own ancestral places and to examine various Kanaka geographies as frameworks for better understanding who we are and where we come from. It is also hoped that this book will have a broad appeal to other peoples who share similar histories and relationships with the ʻāina.

This book honors the moʻolelo of the ancestral places of Kānaka and the relationships that we share with our environment. It is an intensely personal view of how Kanaka geographies relate to place, time, ancestry, and history frame a Kanaka worldview and sense of place. *Ancestral Places* elucidates a Kanaka geography by quoting our kūpuna through their own moʻolelo and by commemorating the ways in which they express their connections to their places. It introduces the reader to the ways in which Kānaka relate to the ʻāina. Chapter 1 introduces the reader to select mele koʻihonua (cosmogonic genealogies) and discusses the importance of knowing one's genealogy. Chapter 2 explores the construction of ancestral Kanaka identities and how these identities are shaped by one's rank in society. Chapter 3 delves into the fluidity of place and how Kānaka transform spaces into personalized places by naming the heavenscapes, landscapes, and oceanscapes. Chapter 4 reveals some of the ancestral cartographic performance methods that Kānaka used to "map" their ancestral places and to retain their moʻolelo. Chapter 5 proposes that over time, Kānaka developed a capacity to receive and perceive stimuli from our environment and to respond to these sensory stimuli in ways that contribute to our overall collective understanding of our world.

During the time that I have been writing this book, I have mourned the passing of two fathers, Frank G. Oliveira and Jacob K. Barros Jr., and I have experienced two births. The first of these births occurred the day my precious daughter, Kahakai K. M. C. M. Apo, was born. The second of these births occurred the day the final draft of this book manuscript was written. This work memorializes the tremendous knowledge of my kūpuna and mākua who have passed and serves as a legacy for the generations of kama yet to be born.

Use of ʻŌlelo Hawaiʻi

I have made a conscious decision to honor the voice of my kūpuna as a mainstream language by not italicizing ʻōlelo Hawaiʻi words. Like Noenoe Silva, "I have not italicized Hawaiian words in the text in keeping with the recent movement to resist making the native tongue appear foreign in writing produced in and about a native land and people."[4] I look forward to the day when there is a critical mass of ʻōlelo Hawaiʻi speakers so that I will be able to write in ʻōlelo Hawaiʻi and italicize the occasional English word to denote it as a foreign language. Until such time becomes a reality, it is important to me that the content of this book be accessible to a variety of audiences in Hawaiʻi and abroad.

In recognition of the fact that many of the readers of this book are not fluent ʻōlelo Hawaiʻi speakers, ʻōlelo Hawaiʻi words are defined the first time they are used in the text, and a glossary appears as an appendix. All translations are mine, except where otherwise noted.

To further assist readers in pronouncing ʻōlelo Hawaiʻi words and names, I have included ʻokina (glottal stops) and kahakō (macrons) throughout the book, except where I quote directly from sources where ʻōlelo Hawaiʻi orthography is absent or where I am unsure about the "correct" pronunciation of proper nouns (e.g., aliʻi [chief] names, place names). In cases where more than one spelling of a proper noun may exist, such as the name of the island Molokaʻi (also known as Molokai) or the name of the aliʻi Kupulanakēhau (spelled Kūpūlanakehau in some texts and Kapulanakēhau in others), I have made my own educated decision about how to spell the name for the purposes of this book and I have noted the alternative spellings that I am aware of.

When writing about ʻōlelo noʻeau, I have elected to regularize ʻōlelo Hawaiʻi spellings to conform with contemporary conventions. At times, contemporary conventions might differ slightly from the way the text appears in the book *ʻŌlelo Noʻeau.* Although I believe that it is generally good practice to quote a text exactly as the author wrote it, I have made the exception here for a couple of reasons. First, Mary Kawena Pukui,[5] author of *ʻŌlelo Noʻeau* and coauthor of the *Hawaiian Dictionary*, herself varied the spelling of some words in these two texts. Second, ʻōlelo noʻeau are wise poetic sayings that many fluent ʻōlelo Hawaiʻi speakers grew up learning and using. Mary Kawena Pukui's book, *ʻŌlelo Noʻeau*, records this oral tradition in written form; therefore, she took her own liberties to document the voices of her informants and reduce their spoken words to the written page. For consistency and clarity, whenever more than one spelling exists, I have opted to use the standardized system currently used today.

While using standardized ʻōlelo Hawaiʻi is helpful to the nonfluent ʻōlelo Hawaiʻi speaker, it is important to also acknowledge that Kānaka highly value ambiguity. ʻŌlelo Hawaiʻi poets, for example, play upon the varied pronunciations and kaona (concealed meanings) of words to convey multiple ideas.[6] Historians openly debate the authenticity of one another's moʻolelo and moʻokūʻauhau (genealogies). ʻŌlelo Hawaiʻi terms sometimes even vary from place to place. For example, a particular type of sea urchin is known as hāʻueʻue in some localities and hāʻukeʻuke in others.

This book celebrates Kanaka geographies and acknowledges the variation in local practices. As a Kanaka who descends from Maui aliʻi, I have written this book from a largely Maui-centric perspective. Whenever possible, Maui examples are used to convey my point. This book was written to honor the legacy of my kūpuna and to ensure that the future generations of kama remember the moʻolelo bequeathed to them.

Chapter 1 Mele Koʻihonua

Moʻolelo (historical accounts), especially those cosmogonic in nature, form the foundation for a Kanaka (Native Hawaiian) geography, illuminating the genealogical connection that Kānaka share with the ʻāina (land; that which feeds). Mele koʻihonua (cosmogonic genealogies) are crucial to understanding a Kanaka worldview, and through these cosmogonic genealogies we learn of the formation of the ʻāina, the first living organisms, and the birth of the akua (gods) and the people. These oral traditions are historical accounts that provide modern scholars with insights regarding ancestral culture, thereby revealing the connection that Kānaka living in ancestral times had with their environment. Through these moʻolelo, relationships are established, described, and reinscribed between the land, ocean, and sky; akua and aliʻi (chiefs); and aliʻi and makaʻāinana (general population).

The moʻolelo of Kānaka commences at the beginning of space and time with cosmogonic genealogies. Mele koʻihonua run the gamut thematically from evolution, birth of islands via the mating of gods, and volcanic eruptions to biblically influenced stories. Joseph Mokuʻōhai Poepoe, editor of the daily ʻōlelo Hawaiʻi newspaper, *Ka Naʻi Aupuni,* wrote as follows in 1906:

> He ekolu no mau mahele nui i ku ai na hoike ana mai a keia mau mookuauhau no ka loaa ana mai o neia mea, he honua a he aina hoi: (1) Ua hanau maoli ia mai no ka mole o ka honua e ka wahine; (2) ua hana lima maoli ia ka honua e ke kanaka; (3) ua ulu a ua puka mai ka aina mailoko mai o ka lipolipo o ka pouli, oia hoi, ka Po, aole mamuli o ka hanauia ana e ka wahine, a hana maoli ia ana paha e ka lima o ke kanaka. (There are three main categories of creation stories: (1) The

taproot of the earth was birthed by a woman; (2) the earth was created by the hands of a person; (3) the land grew and emerged from the depths of darkness, that is from the Pō, not because of being born to a woman or being created by the hand of a person.)[1]

Because there are so many varying accounts, it comes as little surprise that a newspaper article entitled, "Moolelo Hawaii: Mokuna 1: No ka Aina ana ma Hawaii nei," appearing in *Ka Hoku o Hawaii* on December 21, 1911, noted, "He mea kahaha loa no ka manao i ka lohe ana mai i na olelo a ka poe kahiko, no ke ku mua o ka aina ana ma Hawaii nei, he kuee ko lakou manao, aole he like pu" (It is astonishing to hear the theories of the people of old about the formation of the land in Hawai'i; they are contradictory; they are not consistent). Indeed, many contradictory versions of Kanaka creation histories exist, suggesting that Kānaka were open to multiple interpretations of mo'olelo.

In spite of these contradictory accounts, a common element shared by many mele ko'ihonua is the genealogical relationship between the land, humankind, and the gods. Similar to the Māori concept of wairua (spirit), where all living and nonliving elements are believed to be interrelated and possess a spirit, many Kānaka likewise value mele ko'ihonua as the framework by which all things in the natural environment, including people, are genealogically linked and ordered. This chapter explores select mele ko'ihonua, revealing how some accounts are complementary, overlapping with other creation histories, while others are stand-alone accounts differing completely from their counterparts.

While my narrative describes and analyzes Kanaka geographies, to discuss all known mele ko'ihonua is beyond the scope of this book. This chapter cannot be truly exhaustive of the subject, because a multiple-volume series would likely be necessary. The aim is to simply introduce the reader to some of the better-known mele ko'ihonua, because no discussion of Kanaka geography would be complete without providing such an overview.

Kumulipo

Of all Kanaka cosmogonic genealogies, the *Kumulipo* is the best-known today.[2] "Kumu" means "origin, source, foundation," and "lipo" means "dark, night, chaos."[3] The union of these two words denotes the very beginning of time, when only darkness and chaos prevailed. The *Kumulipo* is a story both of origin and evolution, with allusions to the natural growth of a baby within the womb.[4] In the *Kumulipo*, the ʻāina is not born in a natural birth process, nor is it created by the hands of the akua; rather, it grows from the depths of darkness and evolves into ka pae ʻāina Hawaiʻi (the Hawaiian archipelago).[5]

The *Kumulipo* exemplifies how mele koʻihonua avowed the birthright of aliʻi to rule. Composed by Keāulumoku for Kaʻīimamao circa 1700, the *Kumulipo* exalts Kaʻīimamao's high-ranking lineage by tracing his genealogy back to the creation of the world and the gods.[6] As only the second monarch of the Kingdom of Hawaiʻi to be elected to the throne, Kalākaua later used the *Kumulipo* to affirm his royal lineage and pave the way for his sister, Liliʻuokalani to reign as queen after the death of her beloved brother, Leleiōhoku.

Kalākaua and Liliʻuokalani, Kaʻīimamao's great-great-grandchildren, popularized the *Kumulipo*. The pair dramatically increased the general public's access to the mele koʻihonua when Kalākaua printed the *Kumulipo* in 1889.[7] Then in 1895, Liliʻuokalani began translating the mele koʻihonua while she was imprisoned for a period of eight months in her own palace.[8] Translating the *Kumulipo* between 1895 and 1897[9] into English, the language of the colonizer, was a form of political resistance. In an act of defiance against her captors, Liliʻuokalani used English to reclaim her rights as a sovereign. Like her brother before her, Liliʻuokalani turned to her moʻokūʻauhau (genealogy) and mele koʻihonua to reaffirm her birthright to the throne.

The *Kumulipo* was a great source of mana (spiritual power) for those to whom this genealogy belonged. This mele koʻihonua, more than two thousand lines long when reduced to writing,[10] was recited by master genealogists at sacred ceremonies, two of which were noted by Liliʻuokalani. First, Hewahewa along with Ahukai chanted the *Kumulipo* at Koko, Oʻahu, to Alapaʻiwahine at the time that Keʻeaumoku was near death.[11] It was also chanted by Pūʻou, a high-ranking kahuna (priest), at Hikiau Heiau (a temple) in Kealakekua, Hawaiʻi, at a ceremony that included Captain Cook.[12] In recent times, Kanaka practitioners have begun reviving this tradition. In 2003, for example, the *Kumulipo* was recited on the steps of the ʻIolani Palace in honor of Kalākaua's 167th birthday.

Divided into sixteen distinct wā (periods of time), the *Kumulipo* is comprised of seven wā of darkness, followed by nine of light. It begins in the first wā with void, chaos, and deep darkness. It is a time of the spirits. Kumulipo is born in the darkness as a male, and Pōʻele is born in the darkness as a female. Sea creatures are born next, followed in the second wā by the birth of fishes and shrubs. In wā three, insects and birds are born. Reptiles, along with more insects and shrubs, are born in the fourth wā. In wā five and six, pigs and rats are born, respectively. In the seventh and final wā of darkness, the dog makes its appearance.[13]

Finally, after seven wā of night, day is born in wā eight, as is the first human, Laʻilaʻi, a woman. She descends from ancestors of darkness, yet she emerges in the first wā of light. Laʻilaʻi, the ancestor of gods and humans, is born in the same wā as the first man, Kiʻi, and the gods, Kāne and Kanaloa. In the ninth wā, the earth is born, along with several of Laʻilaʻi's children.[14] In the tenth wā, Laʻilaʻi returns to mate with Kāne, while in the eleventh, numerous husband and wife pairs are listed, some of which are gods. Wā twelve is the period when Palikū, Olōlo (also known as Lolo), Kumuhonua, and Hāloa are born.[15]

In Liliʻuokalani's version, she does not list a thirteenth wā; rather, a second branch originating from Palikū is given.[16] In anthropologist Martha Beckwith's *The Kumulipo: A Hawaiian Creation Chant,* this second branch serves as the thirteenth wā.[17] In the fourteenth wā, Kupulanakēhau (also known as Kapulanakēhau)[18] is born as a woman and enters into a relationship with Kahikoluamea to beget Wākea.[19] In wā fifteen, ʻUlu, Nānāʻulu, and the Māui brothers are born.[20] The demigod Māui is a descendant of the ʻUlu line.[21] The last wā lists genealogical pairs down to Piʻilani, aliʻi nui (high-ranking chief) of Maui, and Lāʻielohelohe, then finally to Lonoikamakahiki, also known as Kaʻīimamao.[22]

The *Kumulipo* is essential for gaining an understanding of a Kanaka worldview as well as a Kanaka geography. It reveals several recurring themes, including the struggle for survival and the importance of maintaining pono (harmony, balance) in the world. Harmony is achieved when darkness is balanced with light and male is balanced with female. Pono is also exemplified by the pairing of ocean and land creatures that solidifies the bond between the land and the sea and their interdependence with one another. The *Kumulipo* is also a history of interrelatedness—all plants, animals, kānaka, and akua are genealogically connected.

The *Kumulipo*'s history of interrelatedness extends to other mele koʻihonua as well. Of all the mele koʻihonua, the *Kumulipo* is arguably the most important of the cosmogonic traditions known today because it is often considered to be the most encompassing. Many of the other creation accounts that have been passed down can be found within the *Kumulipo*. That is to say, mele ko'ihonua such as *Kapōhihihi* (branching out of night or chaos), *Kumuhonua* (beginning of the earth), *Olōlo* (brains or oily coconut meat), *Palikū* (vertical precipice), and *Puanue* (the rainbow) are sometimes presented as individual creation stories, when in fact such accounts also appear as sections of the *Kumulipo*.[23]

Papa and Wākea

The mele ko'ihonua of Papa,[24] earth mother, and Wākea,[25] sky father, is one of the many mele ko'ihonua whose genesis originates in the *Kumulipo*. Kahikoluamea and Kupulanakēhau are acknowledged as the parents of Wākea in the twelfth and fourteenth wā of the *Kumulipo*.[26] Papa and Wākea are half brother and sister through the 'Ōpukahonua lineage.[27] Their ancestors were from a distant land known as Kahiki, but the couple settled down in Loloimehani.[28]

Papa and Wākea are a primal pair. As journalist and author Abraham Fornander asserts, "[Ma] ka moolelo o Wakea, ua olelo nui ia, oia na kupuna mua o keia mau aina, a ma o laua la i laha mai ai na kanaka, a o laua na kupuna alii o keia noho ana. Ua oleloia ma ko Wakea mookuaahau [sic] laua a me kana wahine o Papa, ua hanau mai keia mau moku mai loko mai o laua"[29] (In the tradition of Wākea it has been often stated that they were the ancestors of these islands, and that it was through them that the people were born, and they are the ancestors of the chiefs of these islands. It is said in the genealogy of Wākea and his partner Papa that these islands were born to them). In the twelfth wā of the *Kumulipo*, Wākea procreates with Haumea, a manifestation or kino lau of Papa,[30] and his daughter, Ho'ohokukalani (also known as Haohokakalani and Ho'ohōkūkalani).[31] Hāloanaka is born to Ho'ohokukalani as a premature fetus, and a kalo (taro) grows from the place where the fetus is planted.[32] A second child, Hāloa, is born as the kaikaina (younger sibling) of the kalo; this child becomes the first ali'i and kupuna (ancestor) of the Kānaka.[33] The genealogical relationship between the Kānaka, kalo, and 'āina (from which the kalo grows) is revealed in the historical account of Papa and Wākea and their descendants.

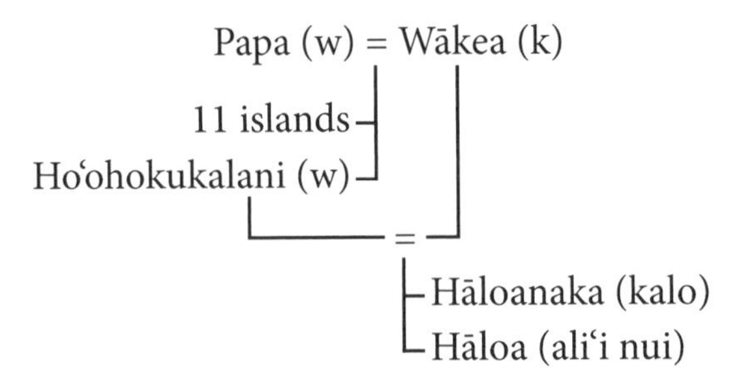

FIG. 1.1 Mo'okū'auhau of Papa and Wākea's offspring.[34]

As implied by her name, Papahānaumoku (Papa that gives birth to islands), Papa birthed some of the islands. Moreover, she is said to have birthed the islands using various parts of her body, from her head to her feet.[35] According to Fornander, "Ma ka moolelo nae o Wakea laua me kana wahine me Papa, i hanau maoliia mai keia mau aina mai loko mai o laua. O Hawaii ke keiki mua a Papa laua me Wakea, a mahope hanau mai o Maui, a pela i hanau ai a he umikumamakahi moku,[36] a o Kahoolawe ka moku aole i pili i loko o Wakea laua me Papa"[37] (In the historical account of Wākea and his partner Papa, these islands were born to them. Hawai'i was the first child of Papa and Wākea, and Maui was born afterwards, and in this manner eleven islands were born. And as for Kaho'olawe, the island, it was not born to Papa and Wākea). Nevertheless, Poepoe claims that Papa was not the biological mother of the islands, and that historians misinterpreted compositions. Poepoe argues that such mele (songs) credit Papa with birthing famous *descendants* of islands from Hawai'i to Kaua'i, rather than the islands themselves. Instead of identifying these descendants by their given names, reference is simply made to their island of residence.[38]

Composed by Pāku'i, a renowned historian and contemporary of Kamehameha, *Mele a Pāku'i* is a mo'olelo and mele that enumerates the birthing of ka pae 'āina Hawai'i.[39] In *Mele a Pāku'i,* Papa and Wākea are the parents of Kahitikū, Kahitimoe, Ke'āpapanu'u,[40] Ke'āpapalani, and Hawai'i.[41] Papa then gives birth to the island of Maui(loa), whose paternity varies depending on the version of *Mele a Pāku'i* cited. According to Poepoe, "O Wakea la ua kane" (Wākea is the aforementioned man), but according to Fornander, Maui is born to "Wakea laua o Kane" (Wākea and Kāne).[42] Per Poepoe, "He nuu no Ololani, no Lono, no Ku, o Kane ma laua o Kanaloa" ([Mauiloa is] a high-ranking one for Ololani [an acclaimed chief], for Lono, for Kū, Kāne, and Kanaloa), while Fornander's version states, "He nui Mololani no Ku, no Lono, o Kane ma laua o Kanaloa" (Mololani [well-kept one] is of great importance, for Kū, for Lono, Kāne, and Kanaloa).[43] Papa then travels to Kahiti. In Papa's absence, Wākea has a relationship with Kāulawahine, and Lāna'ikāula, Kāulawahine's eldest child, is born. Next, Wākea procreates with Hina, and Moloka'iahina (also known as Molokaiahina) is born. When Papa returns to ka pae 'āina Hawai'i, she is angered that Wākea has taken other lovers, so she too takes a new lover, Lua, and O'ahualua is born. Finally, Papa returns to Wākea, and Kaua'i is born, followed by the islands of Ni'ihau, Lehua, and Ka'ula.[44]

Mele a Pakui

1.	O Wakea a Kahiko Luamea, a—e,		O Wakea Kahiko Luamea,
2.	O Papa, o Papa-hanau-moku ka wahine,		O Papa, o Papahanaumoku ka wahine,
3.	Hanau o Kahiki ku, Kahiki-moe,		Hanau Tahiti-ku, Tahiti-moe,
4.	Hanau ke apaapaa nuu, ke apaapaa lani		Hanau Keapapanui,
		5	Hanau Keapapalani,
5.	Hanau *Hawaii* ka moku hiapo.		Hanau Hawaii;
			Ka moku makahiapo,
6.	He keiki makahiapo a laua—a—a.		Keiki makahiapo a laua.
7.	O Wakea la ua kane,		O Wakea laua o Kane,
8.	O Papa, o Walinuu ka wahine,	10	O Papa o Walinuu ka wahine.
9.	Hookauhua Papa i ka moku,		Hookauhua Papa i ka moku,
10.	Ho-iloli ia *Maui*;		Hoiloli ia Maui,
11.	Hanau Maui-loa, he moku,		Hanau Mauiloa he moku;
12.	I hanauia he ololani, he uilani,		I hanauia he alo lani,
13.	Uilani he-i kapa lau maewa	15	He Uilani-uilani,
			Hei kapa lau maewa.
14.	He nuu no Ololani, no Lono, no Ku,		He nui Mololani no Ku, no Lono,
15.	No Kane ma laua o Kanaloa—o—a,		No Kane ma laua o Kanaloa.
16.	Hanau kapu ke kuakoko		Hanau kapu ke kuakoko,
17.	Kaahea Papa ia Kanaloa, he moku	20	Kaahea Papa ia Kanaloa, he moku,
18.	I hanauia he pu-nua, he naia,		I hanauia he punua he naia,
19.	He keiki i'a na Papa i hanau		He keiki ia na Papa i hanau,

20. Haalele Papa hoi i Kahiki		Haalele Papa hoi i Tahiti,
21. Hoi a Kahiki Kapakapaka-ua		Hoi a Tahiti Kapakapakaua.
22. Moe o Wakea—	25	Moe o Wakea moe ia Kaulawahine
23. Moe ia Kaula-wahine		
24. Hanau *Lanai* a Kaula,		Hanau o Lanai Kaula.
25. He keiki makahiapo na ia wahine;		He makahiapo na ia wahine.
26. Hoi Wakea loaa Hina		Hoi ae o Wakea loaa Hina,
27. Loaa Hina, he wahine moe na Wakea		Loaa Hina he wahine moe na Wakea,
28. Hapai Hina ia *Molokai* he moku,	30	Hapai Hina ia Molokai, he moku,
29. O Molokai a Hina he keiki moku,		O Molokai a Hina ke keiki moku.
30. Haina e ke kolea a Laukaula		Haina e ke kolea o Laukaula
31. Ua moe o Wakea i ka wahine		Ua moe o Wakea i ka wahine.
32. Ena Kalani, ku ka hau lili o Papa—a—pa,		O ena kalani, kukahaulili o Papa.
33. Hoi mai o Papa mailoko mai o Kahiki-ku	35	Hoi mai Papa mai loko mai o Tahiti;
34. Ku inaina, lili i ka punalua		Inaina lili i ka punalua;
35. Hae manawa ino i ke kane o Wakea		Hae, manawaino i ke kane, o Wakea,
36. Moe ia Lua, he kane hou ia		Moe ia Lua he kane hou ia.
37. Hanau o *Oahu* a Lua;		Hanau Oahu-a-Lua,
38. Oahu a Lua, he keiki moku	40	Oahu-a-Lua, ke keiki moku,
39. He keiki maka-na-lau na Lua—u—a,		He keiki makana lau na Lua.
40. Hoi hou aku no noho me Wakea		Hoi hou aku no moe me Wakea.
41. Naku Papa i ka moku o Kauai		Naku Papa i ka iloli,
		Hoohapuu Papa i ka moku o Kauai
42. Hanau Kamawaelualani, he moku	45	Hanau Kamawaelualanimoku,
43. He wewe Niihau, he palena Lehua,		He eweewe Niihau;
		He palena o Lehua,
44. He panina Kaula o ka Moku Papapa.		He panina Kaula.
		O ka Mokupapapa.
		(continues)

1. Wākea Kahiko Luamea
2. Papa, Papa-hānau-moku, the woman,
3. Born is Kahiki kū, Kahiki-moe,[45]
4. Born is Keʻāpaʻapaʻanuʻu, Keʻāpaʻapaʻalani[46]
5. *Hawaiʻi* is born as the eldest island.
6. Their eldest child.
7. Wākea is the aforementioned male,
8. Papa of Walinuʻu (or Papa also known as Walinuʻu) is the female,
9. Papa experiences pregnancy sickness due to the island,
10. Suffering pregnancy pains with *Maui*;
11. Maui-loa is born as an island,
12. Born an acclaimed chief, a chiefly beauty,
13. A chiefly beauty ensnared in the swaying kapa
14. A high-ranking one for Ololani, for Lono, for Kū,
15. For Kāne folks and Kanaloa,
16. Consecrated are the birth pains
17. Papa suffers with Kanaloa, an island
18. Born a fledgling, a dolphin,
19. A fish child born to Papa
20. Papa left, returned to Kahiki
21. Returned to Kahiki Kapakapaka-ua
22. Wākea slept
23. with Kāulawahine
24. *Lānaʻi* a Kāula was born,
25. An eldest child of this woman;
26. Wākea returns to Hina

Wākea Kahiko Luamea,
Papa, Papahānaumoku, the woman,
Born is Tahiti-kū, Tahiti-moe,
5 Born is Keʻāpapanui, Keʻāpapalani
Hawaiʻi is born;
The eldest island,
Their eldest child.
Wākea and Kāne are the males,
10 Papa of Walinuʻu (or Papa also known as Walinuʻu) is the female.
Papa experiences pregnancy sickness due to the island,
Suffering pregnancy pains with Maui,
Mauiloa is born as an island;
Born in chiefly presence,[47]
15 A chiefly beauty, chiefly beauty,
Ensnared in the swaying kapa.
Mololani is of great importance, for Kū, for Lono,[48]
For Kāne folks and Kanaloa.
Consecrated are the birth pains,
20 Papa suffers with Kanaloa, an island,
Born a fledgling, a dolphin,
A fish child born to Papa,
Papa left, returned to Tahiti,
Returned to Tahiti Kapakapakaua.
25 Wākea slept with Kāulawahine
Lānaʻi Kāula was born,
An eldest child of this woman.
Wākea returns to Hina

27. Hina is begotten as a female companion for Wākea
28. Hina is pregnant with *Molokaʻi*, an island,
29. Molokaʻi a Hina is an island child,
30. It is told by the plover, Laukaula
31. Wākea slept with the woman
32. Papa rages with anger and jealousy,
33. Papa returns from Kahiki-kū
34. Hatred toward the other lover
35. Wild with rage at the man, Wākea
36. Slept with Lua, a new man
37. *Oʻahu* a Lua is born;
38. Oʻahu a Lua, an island child
39. A leaf opening child for Lua,
40. Returns to sleep with Wākea
41. Papa suffers birth pains with the island of Kauaʻi
42. Kamāwaelualani is born, an island
43. Niʻihau, a sprouting lineage, Lehua is a border,
44. Kaʻula is the closing one of the Moku Papapa.

SOURCE: Poepoe, "Ka Moolelo Hawaii Kahiko: Mokuna I: Na Kuauhau Kahiko e Hoike ana i na Kumu i Loaa ai ka Pae Moku o Hawaii nei," *Ka Naʻi Aupuni*, February 2–3, 1906.

Hina is begotten as a female
companion for Wākea
30 Hina is pregnant with Molokaʻi, an
island
Molokaʻi a Hina is an island child.

It is told by the plover, Laukaula
Wākea slept with the woman.
Papa rages with anger and jealousy.

35 Papa returns from Tahiti;
Hatred toward the other lover;
Wild with rage at the man, Wākea,

Slept with Lua, a new man.
Oʻahu a lua is born,
40 Oʻahu a lua, the island child,
A leaf opening child for Lua.
Returns to sleep with Wākea.
Papa suffers birth pains,
Papa gives birth to the island of
Kauaʻi
45 Kamāwaelualanimoku is born,

Niʻihau, a sprouting lineage;
Lehua is a border,
Kaʻula is the closing one
Of the Mokupapapa.
(continues)

SOURCE: Fornander, *Fornander Collection*, vol. 4, 12–15.

Wakea a Kahiko Luamea (k) = Papa-hanau-moku (w)

Kahiki ku
Kahiki-moe
ke apaapaa nuu
ke apaapaa lani
Hawaii
Maui(loa)

Papa
Kanaloa

Papa travels to Kahiki:

Kaula-wahine (w) = Wakea (k) = Hina (w)

Lanai a Kaula   Molokai a Hina

Papa returns from Kahiki-ku:

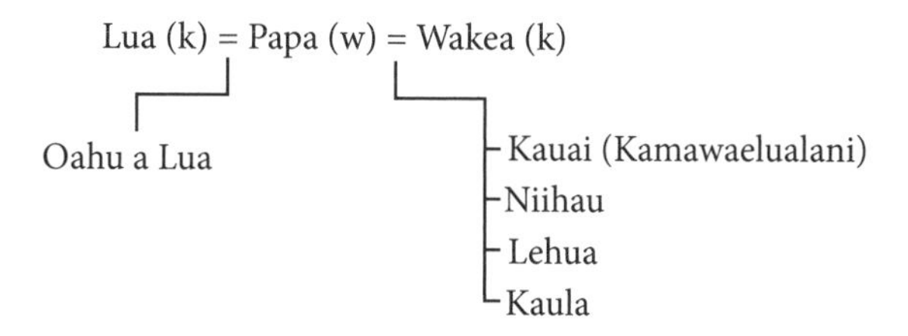

FIG. 1.2. Diagram of *Mele a Pakui* per Poepoe

SOURCE: Poepoe, "Ka Moolelo Hawaii Kahiko: Mokuna I: Na Kuauhau Kahiko e Hoike ana i na Kumu i Loaa ai ka Pae Moku o Hawaii nei," *Ka Na'i Aupuni*, February 2–3, 1906.

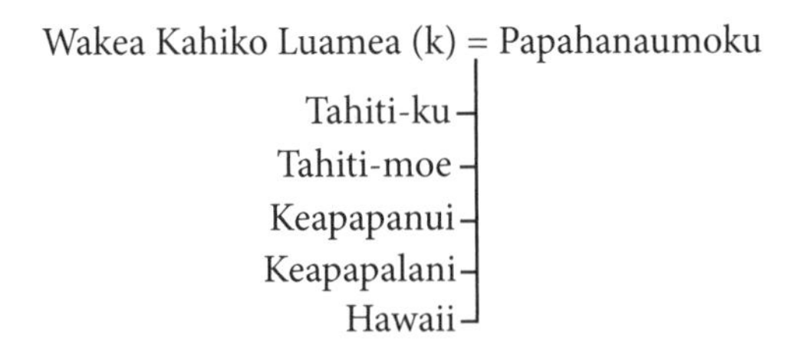

Kane (k)
Wakea (k) = Papa (w)

Maui(loa)

Papa (w)

Kanaloa

Papa travels to Tahiti:

Kaulawahine (w) = Wakea (k) = Hina (w)

Lanai Kaula

Molokai a Hina

Papa returns from Tahiti:

Lua (k) = Papa (w) = Wakea (k)

Oahu-a-Lua

Kauai (Kamawaelualanimoku)

Niihau

Lehua

Kaula

FIG. 1.3. Diagram of *Mele a Pakui* per Fornander

SOURCE: Fornander, *Fornander Collection*, vol. 4, 12–15.

In another Papa and Wākea mele ko'ihonua, *'O Wākea Noho iā Papa-hānaumoku,* Papa is the mother of the islands Hawai'i and Maui. Ho'ohokukalani gives birth to Moloka'i and Lāna'i, and Papa becomes jealous of her daughter, Ho'ohokukalani, for having sexual relations with Wākea. Papa returns to Wākea and gives birth to the islands of O'ahu, Kaua'i, and Ni'ihau.[49]

O Wakea noho ia Papa-hanau-moku,	Wākea lived with Papa who gives birth to islands,
Hanau o Hawaii, he moku,	Born is Hawai'i, an island,
Hanau o Maui, he moku.	Born is Maui, an island.
Hoi hou o Wakea noho ia Hoo-hoku-ka-lani.	Wākea returns to live with Ho'ohokukalani.
Hanau o Molokai, he moku,	Born is Moloka'i, an island,
Hanau o Lanai ka ula, he moku.	Born is Lāna'i ka 'ula, an island.
Lili-opu-punalua o Papa ia Hoo-hoku-ka-lani.	The womb of Papa is jealous of her partner Ho'ohokukalani.
Hoi hou o Papa noho ia Wakea.	Papa returns and lives with Wākea.
Hanau, o Oahu, he moku,	Born is O'ahu, an island,
Hanau o Kauai, he moku,	Born is Kaua'i, an island,
Hanau o Niihau, he moku,	Born is Ni'ihau, an island,
He ula a o Kahoolawe.	Kaho'olawe is a red rock.

SOURCE: Nathaniel B. Emerson's Notes in *Hawaiian Antiquities*, 243.

Papa's birthing of the islands is but one of the many Kanaka origin accounts. In some traditions, it was actually Wākea who formed the islands with his bare hands.[50] Another tradition claims that Papa gave birth to an ipu (gourd; calabash), the cover of which was then flung upwards, forming the heavens. The flesh and seeds of the ipu became the sky—complete with the sun, moon, and stars. Rain was made from the juice of the ipu, while the land and sea were made with the body of the ipu.[51]

Mele a Kamahualele

Mele a Kamahualele refutes Papa and Wākea as being the original progenitors of ka pae moku (the Hawaiian archipelago). According to this tradition,

> I ka manawa i holo mai ai o Moikeha mai Tahiti mai, mamuli o ka hoaaia i kana wahine manuahi ia Luukia, no ko Mua olelo hoopunipuni ana ia Luukia no ka hewa i hana oleia e Moikeha, aka ma kela lohe ana o Moikeha ua hana pono ole ia oia, nolaila, haalele oia ia Tahiti, holo mai oia i Hawaii nei, a i ka hookokoke ana mai o na waa e pae i Hilo, ia manawa, ku mai o Kamahualele i luna o ka pola o na waa, a kahea mai: (At the time that Mo'ikeha sailed from Tahiti because his lover, Lu'ukia, had become outraged by Mua's false accusations of Mo'ikeha's infidelity; therefore, Mo'ikeha left Tahiti and sailed to Hawai'i, and as the canoes neared the shores in Hilo, Kamahualele stood on the cross-boards of the canoe and chanted:)[52]

Eia Hawaii, he moku, he kanaka,	Behold Hawai'i, an island, a man,
He Kanaka Hawaii-e.	A Kanaka Hawaii.
He Kanaka Hawaii,	A Kanaka Hawaii,
He Kama na Tahiti,	A child of Tahiti,
He Pua Alii mai Kapaahu.	A royal descendant from Kapa'ahu.
Mai Moaulanuiakea Kanaloa,	From Moa'ulanuiākea Kanaloa,
He Moopuna na Kahiko laua o Kapulanakehau.	A descendant of Kahiko and Kapulanakēhau.
Na Papa i hanau,	It was Papa that birthed,
Na ke Kama wahine a Kukalaniehu laua me Kahakauakoko.	The daughter of Kūkalani'ehu and Kahakauakoko.
Na pulapula aina i paekahi,	Sprouts of land in a row,
I nonoho like i ka hikina, komohana,	Residing similarly from east to west,
Pae like ka moku i lalani,	Situated evenly in a row,
I hui aku hui mai me Holani.	Gathered to, gathered with Hōlani.
Puni ka moku o Kaialea ke kilo,	Kaialea, the seer, circumnavigated the islands,
Naha Nuuhiwa lele i Polapola:	Nukuhiwa is out of sight; gone to Borabora:
O Kahiko ke kumu aina,	Kahiko is the source of land,
Nana i mahele kaawale na moku,	He divided and separated the islands,

Moku ke aho lawaia a Kahai,	The fishing line of Kaha'i is severed
I okia e Kukanaloa,	Cut by Kūkanaloa,
Pauku na aina, na moku,	The lands, the islands are divided,
Moku i ka ohe kapu a Kanaloa.	Severed by the sacred bamboo of Kanaloa.
O Haumea manu kahikele,	Haumea manu kahikele,
O Moikeha ka lani nana e noho.	Mo'ikeha is the chief who will reside there.
Noho kuu lani ia Hawaii-a-	My beloved chief dwells in Hawai'i
Ola! Ola! O Kalanaola.	Live! Live! Kalanaola.
Ola ke alii, ke kahuna.	Long live the chief, the priest.
Ola ke kilo, ke kauwa;	Long live the seer, the servant;
Noho ia Hawaii a lulana,	They shall reside calmly in Hawai'i,
A kani moopuna i Kauai.	There shall be descendants on Kaua'i.
O Kauai ka moku-a-	Kaua'i, the island
O Moikeha ke alii.	Mo'ikeha is the chief.

SOURCE: Fornander, *Fornander Collection,* vol. 4, 21

Kamahualele—a well-respected prophet and historian who is credited with chanting *Mele a Kamahualele* in honor of the arrival of his ali'i, Mo'ikeha—suggests that the progenitors of ka pae 'āina Hawai'i sailed to these islands from Tahiti. Therefore, Papa and Wākea were not the progenitors of Kānaka.[53]

Kumulipo-Kumuhonua

Although Papa and Wākea are considered to be a primal pair, Laka—son of Kumuhonua and Lalohonua—precedes Papa and Wākea by thirty-six generations in the *Kumulipo*.[54] Together, the *Kumulipo* and *Kumuhonua* mele ko'ihonua are often called *Ka Mo'olelo Kumulipo-Kumuhonua o Hawai'i* (The Account of Kumulipo-Kumuhonua of Hawai'i) or *Ka Mo'olelo Hawai'i Kahiko* (The History of Ancient Hawai'i) because these genealogies overlap.[55] In the *Kumuhonua* genealogy, it is generally believed that Kumuhonua (also known as Honua'ula,[56] Hulihana,[57] Hulihonua,[58] Kānelā'auuli,[59] Kuluipo,[60] and Kumuuli[61]) was the first man.[62]

Although Samuel Kamakau,[63] a highly respected Kanaka scholar and historian, states that some credit Kamaieli (also known as Hālōihoilalo) with giving birth to the mole (taproot, foundation of an island) and that her husband was Kumuhonua,[64] in another version, Kumuhonua, the first man, was created from muddy water. The first woman, Lalohonua, was made from his side. Similar to the *Kumulipo,* this mele ko'ihonua contends that there was a period of darkness. Unlike the seven wā outlined in the *Kumulipo,* however, there are four phases in the *Kumuhonua:* Pōloa, Pōnuiauwaea, Pōkanaka, and Pōhana. Fornander asserts that the god Kāne (also known as Kāneikapōloa) lived during the period of Pōloa, the time of darkness in which there was no heaven or earth. During the time known as Pōnuiauwaea, the gods Kāne, Kū, and Lono created the world. It was also during this phase that light broke for the first time. In Pōkanaka, the first man, Kumuhonua, was created. Before the god Kāne ascended to the heavens, he declared his mandate regarding a forbidden tree. While Kumuhonua resided on the earth, he broke the mandate of Kāne and was punished. The fourth phase, Pōhana, encompasses the generations from Kumuhonua to Nu'u. The second part of Pōhana is Pōauhulihia. During Pōauhulihia, a massive flood struck as punishment for Kumuhonua's disobedience.[65] Nu'u built a canoe named Wa'ahālauali'iokamoku to escape the flood.[66]

Kepelino, a Kanaka scholar and early convert of the Catholic Church, asserts in the historical account of *Kumuhonua* that Kāne was the creator of all things, including night, light, heaven, earth, gods, water, sun, moon, stars, and man.[67] Fornander contends that the account of *Welaahilani* is virtually the same as that of *Kumuhonua,* with the exception of name changes.[68] Similarly, historian James Kaulia's account states that the phase before the creation of the heavens and earth was Kapoliku-Puhohō (referred to later in the same newspaper article as Polikua-Puhohō). This period of darkness was thought to be a female by the name of Kāhulikāhelanuiākeaokonouliʻohupano-panopuhohō. The heavens and earth were created by Kapūkuʻiakua and this female. Kaulia adds that the god Kāne drew the figure of man in the sand and that image became a man named Welaahilaninui. Kaulia further asserts that Welaahilaninui was the "Adamu mua iloko o ka Lahui Hawaii" (first Adam in the Hawaiian Nation). The first woman was created by Kāne using a piece torn from Welaahilaninui's body; her name was Oure (also known as ʻOwē).[69] Similarly, Kamakau credits a supreme god with transforming the god's breath and ipu into the heavens and earth. The god then ordered Kū, Lono, Kāne, and Kanaloa to create man to resemble spirits. Kū, Lono, and Kāne worked together, drawing an image of a man on the sand and transforming the image into a man with a soul.[70] In this tradition, Welaahilaninui was this first man.[71] Kanaloa, unlike the others, was unsuccessful at creating a man from his image. Devastated by his inability to bring life to his image, Kanaloa became enraged and jealous of his counterparts; therefore, it is suggested that Kanaloa is responsible for many of the evil things on earth.[72] The god then commanded them to create a woman. The woman, ʻOwē, was created from a part of Welaahilaninui's body.[73]

The *Kumuhonua* genealogy is important because it demonstrates that as the societal norms changed, so too did the genealogies. With the influence of the missionaries, certain cosmogonic genealogies began to reflect the beliefs of Christianity and were rearranged to support biblicized traditions.[74]

Olōlo/Lolo

In the mele koʻihonua known as *Olōlo* or *Lolo*,[75] Olōlo is born in the twelfth wā of the *Kumulipo*.[76] Welaahilaninui and ʻOwē are the parents of Kahikoluamea, the first man and human.[77] Kamakau suggests that Welaahilaninui and ʻOwē are the original ancestors of the Kānaka;[78] nevertheless, their son Kahikoluamea is honored as the original ancestor in the following *Mele a Kamahualele:* "O Kahiko ke kumu aina / Nana i mahele kaawale na moku" (Kahiko is the source of the land. / The one that divided and separated the islands).[79] Kupulanakēhau is Kahikoluamea's wife.[80] Like Welaahilaninui and

ʻOwē, Kahikoluamea and Kupulanakēhau also resided in Kamāwaelualani.[81] Kahikoluamea and Kupulanakēhau are the parents of Wākea,[82] Līhauʻula (also known as Lihaʻula or Lehuʻula), and Makuʻu.[83]

Wākea is most widely recognized for his relationships with Papa and Hoʻohokukalani. Yet Wākea also played an important role in establishing the Kanaka society, along with his siblings. The descendants of Wākea became aliʻi, while Līhauʻula and Makuʻu were the ancestors of the kāhuna (priests; masters of arts) and makaʻāinana respectively.[84]

ʻŌpuʻukahonua/ʻŌpukahonua

In the mele koʻihonua known as *ʻŌpuʻukahonua* or *ʻŌpukahonua*, ʻŌpuʻukahonua migrated from Tahiti with his wife Lana and his two younger brothers, Lolomu and Mihi, becoming the ancestors of the Kānaka.[85] The *ʻŌpuʻukahonua* genealogy is important because many famous and powerful aliʻi are descendants of ʻŌpuʻukahonua and Lana. Genealogists of the islands of Maui and Hawaiʻi in particular favored this moʻokūʻauhau.[86] Direct descendants of this pair include Kahikoluamea and his wife Kupulanakēhau; Wākea, his wife Papa, and their daughter Hoʻohokukalani; Hāloa, the son of Wākea and Hoʻohokukalani; Waia; ʻUlu; Nānāʻulu; and Akalana and his famous sons, Māuimua, Māuihope, Māuikiʻikiʻi, and Māuiakalana.[87]

In this mele koʻihonua, the ʻāina emerges from the depths of the ocean. According to Fornander,

> Ma ka moolelo hoi o Opuukahonua, ua oleloia, i loaa o Hawaii nei i lawaia ia, a o Opuukahonua ke kupuna mua o ka laha ana o keia lahui. A penei ka olelo a kekahi kanaka kakaolelo o Kahakuikamoana kona inoa, kekahi kanaka kaulana o loko o ka papa kahuna o ka oihanakahuna. (In the historical account of ʻŌpuʻukahonua, it is stated that Hawaiʻi was caught by being fished up, and ʻŌpuʻukahonua was the progenitor of this nation. This is the account of a historian by the name of Kahakuikamoana, a famous priest of the priestly order).[88]

Ea mai Hawaiinuiakea,
Ea mai loko, mai loko mai o ka po.
Puka mai ka moku, ka aina,
Ka lalani aina o Nuumea,
Ka pae aina o i kukulu o Tahiti.
Hanau o Maui he moku, he aina,
Na kama o Kamalalawalu e noho.

Hawaiʻinuiākea arose,
Arising from inside the darkness.
The island, the land, appeared,
The row of islands of Nuʻumea,
The archipelago on the borders of Tahiti.
Maui was born an island, a land,
A dwelling place for the descendants of Kamalālāwalu.

Na Kuluwaiea o Haumea he kane,
Na Hinanuialana he wahine
Loaa Molokai, ke akua, he kahuna,
He pualena no Nuumea,
Ku mai ke alii ka lani.
Ka haluku wai ea o Tahiti.
Loaa Lanai he keiki hookama.
Na Keaukanai i moe aku,
Moe ia Walinuu o Holani,
He *kekea* kapu no Uluhina,
Hanau Kahoolawe, he lopa.
Kiina aku Uluhina
Moku ka piko o ke kamaiki,
Ka iewe o ke keiki i lele
I komo i loko o ka ape nalu,

Kuluwaiea of Haumea the male,
Hinanuialana the female
Molokaʻi was born, a god, a priest,
The dawning light from Nuʻumea,
The chief stands, the heavenly one.
The life-giving water drops of Tahiti.
Lānaʻi was begotten as an adopted child.
It was Keaukanaʻi who married,
Married Walinuʻu from Hōlani
A sacred albino for Uluhina
Kahoʻolawe was born, a foundling.
Uluhina was fetched
The navel of the child was cut,
The afterbirth of the child was thrown
Entering into the folds of the rolling surf,

Ka apeape kai aleale,
Loaa ka malo o ke kama,
O Molokini ka moku
He *iewe* ia-a. He iewe ka moku.

The froth of the raging sea,
The loincloth of the child was gotten,
Molokini the island
A navel string. The island is a navel string.

Ku mai Ahukinialaa,
He alii mai ka nanamu,
Mai ka api o ka ia,
Mai ka ale poi pu o Halehalekalani,

ʻAhukinialaʻa arose,
A chief from the foreign land,
From the gills of the fish,
From the breaking waves of Halehalekalani,

Loaa Oahu, he wohi,
He wohi na Ahukinialaa.
Na Laakapu he kane ia,
Na Laamealaakona he wahine.

Oʻahu was gotten, a wohi,
A wohi from ʻAhukinialaʻa.
From Laʻakapu, a man,
From Laʻamealaʻakona a woman.

Hookauhua, hoiloli i ka Nuupoki alii,	Sickened by child conception and pregnancy carrying the chief Nuʻupoki,
Ka heiau kapu a Nonea	The sacred temple of Nonea
I kauila i ka po kapu o Makalii.	During the lightning in the sacred night of Makaliʻi.
Hanau Kauai he alii, he kama, he pua alii,	Kauaʻi was born a chief, a descendant, a royal progeny,
He huhui alii, a Hawaii,	Of the chiefly order of Hawaiʻi,
Na ke poo kelakela o na moku.	Hawaiʻi the foremost of the islands.
I paholaia e Kalani.	Spread out by Kalani.
Holo wale na moku i Holani,	The ships sail to Hōlani,
I ka wewehi kapu a ka lanakila.	In the sacred grandeur of victory.
Kulia i ka moku a Kanekanaloa,	Standing on the island of Kānekanaloa
Ka ihe laumaki i Polapola.	The barbed spear from Polapola.[89]
Nana i mahiki Wanalia.	That pricked and uplifted Wanalia.[90]
O Wanalia ke kane,	Wanalia, the man,
O Hanalaa ka wahine,	Hanalaʻa, the woman,
Hanau Niihau he aina, he moku,	Niʻihau, a land, an island was born (to them),
He aina i ke *aa* i ka mole o ka aina.	A land at the roots, the taproot of the land.
Ekolu lakou keiki,	There were three children,
I hanau i ka la kahi,	Born on the same day,
O Niihau, o Kaula, Nihoa pau mai,	Niʻihau, Kaʻula, finally Nihoa,
Pa ka makuwahine [sic],	The mother ceased bearing children,
Oili moku ole mai mahope.	No other islands appeared afterwards.
Na Kalani e hoolaa na moku,	It is Kalani who consecrated the islands,
Kau iluna o Nuumea	Perched above Nuʻumea
I ka ahui alii o Kaialea.	Among the royal order of Kaialea.
Na ka lanakila e au na moku.	It is the conqueror who rules the islands.
I huia na kolu e Kalani;	The thirds were united by Kalani;
O Hilo, O Puna, o Kau, lele wale.	Hilo, and Puna, and Kaʻū were thrown in.

Ku mai Kalani me ke kahuna,	Kalani stands with the priest,
Kilohi mai ia Maui a Kama.	Gazing at Maui of Kama.
Aole e u aku puni ka aina	It was not long and the land was circuited
Ke kalele a Kalanimakahakona,	Through the support of Kalanimakahakona,
A ka uiaa i kilakila,	The brave one that commands admiration,
Ke koa nui o lanakila,	The great warrior of victories,
Nana i keehi Oahu.	The one who trampled over O'ahu.
Nakolo na moku i ka pea	The ships sailed rustling due to the sails
I ka maha o Kauai, malia.	To the comfort of Kaua'i, peace.
Puni na aina ia Kalani,	All the islands were circled by Kalani,
Ia Kalanialonoapii,	By Kalanialonoapi'i,
Ke kumu alii o Haloa.	From the chiefly line of Hāloa.
Ea mai Hawaii ka moku;	Sovereign is Hawai'i, the island;
Ea pu me ka lanakila—la.	Rising up with victory.

(aole i pau)

SOURCE: Fornander, *Fornander Collection*, vol. 4, 3, 5.

Like the *Kumulipo* that was written for an ali'i nui, Ka'īimamao, this mele, *Mele a Kahakuikamoana,* appears to have been written for the ali'i nui Kamehameha. After describing the manner in which each of the islands was begotten, the mele references Kamehameha's conquest of ka pae 'āina Hawai'i. With Hawai'i Island already under his control, the mele focuses on how each of the other three major islands (along with the smaller islands associated with these major islands)—namely Maui, O'ahu, and Kaua'i—were conquered.[91]

Puanue

Kumukanikeka'a, wife of Paiaalani, gave birth to the heavens and the corners of the earth in the *Puanue* genealogy.[92] Other mele ko'ihonua insist that Puanue was born hundreds of generations after La'ila'i, the first woman.[93] In the *Kumulipo,* La'ila'i is born in the eighth wā, while Puanue is born in the twelfth; therefore, Puanue is not the beginning of Kānaka.[94]

Conclusion

Many are intrigued by the varying accounts of mele ko'ihonua. Why do so many different accounts exist? Kamakau attributes the discrepancies in contradictory versions to human error; people may have accepted the mo'olelo they heard as true, when in fact some may have been incomplete, only enumerating the most important individuals of a family line.[95] Over time, people may have forgotten the origin of man and just memorized the traditions they heard—assuming they were the oldest. Kānaka suggest that "Ke hi'i lā 'oe i ka paukū waena, he neo ke po'o me ka hi'u" (When you hold onto the middle section [of a genealogy], the head [or beginning] and the tail [or the ending of the genealogy] are forgotten). Both are left wanting.[96] Ancestral Kānaka were also highly metaphorical in their expressions; it is possible that kaona (multiple layers of meaning; hidden meaning) was implicit in some genealogies but was mistakenly accepted at face value, as succeeding generations did not have the same in-depth understanding of the family history as their ancestors did.

Another possible explanation for varying accounts is the fact that mele ko'ihonua were sources of power. For ali'i, being able to trace one's ancestry back to the akua and the very creation of the universe increased their mana. The higher the rank of an ali'i, the more mana an ali'i inherited; thus, mele ko'ihonua were tools wielded by ali'i to validate their royal heritage and right to rule. In the case of Kalākaua, the *Kumulipo* helped to elevate his status in the eyes of his constituents. Amongst one's contemporaries, competing genealogies may have been orally recorded. Those who wished to gain mana advocated for their own genealogy as being the "truth."[97] As Martha Beckwith suggests, "A family chant like the Kumulipo, passed down orally from one generation to the next without the stabilizing force of a written text, must have been constantly exposed to political changes within the family and to the urge felt by a new song-maker to revitalize the old memorial by giving it a fresh application to more recent family events."[98] These kinds of genealogies could make or break an ali'i. Those ali'i with the highest genealogies had the most mana and were revered by the maka'āinana as gods that walked the earth.[99] Thus, genealogists had the great kuleana (responsibility) of memorizing the genealogies of ali'i; the incorrect recitation of a genealogy was at times punishable by death.

Genealogists of various ali'i often disputed the authenticity of other genealogies in an attempt to elevate the genealogy of their own ali'i. Genealogists who found discrepancies and fallacies in others' family lineages were praised by their ali'i. Those who falsified their genealogies were compared to the Kaua'ula wind of Maui in the saying, "Ho'okohu Kaua'ula, ka makani o 'Ulupa'u." This famous line from an ancient chant makes reference to one who steals someone else's genealogy and then brags about illustrious relatives that one is not rightfully entitled to (ON 116).

Those with questionable genealogies were said to be "He ali'i no ka malu kukui" (A chief of the shade of the kukui) (ON 63). Such a remark suggests that the person in question had a "shady" genealogy. Any tree that provides shade could have been selected for this wise saying. The choice of the kukui (candlenut tree) was undoubtedly intentional. The kukui provided light at night; therefore, it is often used metaphorically in reference to light, enlightenment, knowledge, and wisdom. Used in this context, a person with a "shady" background not only lacks a reputable genealogy but also a solid foundation in the way of enlightenment. Anyone with a debatable lineage might be said to have "he pili nakekeke" (a relationship that rattles), meaning that the genealogy one purported to belong to was not a proper fit (ON 96). The fact that so many 'ōlelo no'eau (wise sayings; proverbs) were coined with reference to genealogies demonstrates the importance of mo'okū'auhau—especially those cosmogonic in nature—in ancestral times.

In spite of their numerous inconsistencies, cosmogonic genealogies are useful for modern scholars in that they give us an idea of the sequence of events that occurred in ancestral times. These oral traditions of the past should not be dismissed as merely legends and myths; rather, they are historical accounts that provide modern scholars with insights regarding ancestral-culture, revealing the connection that Kānaka living in pre-contact times had with their environment. Such histories enable us to better understand our ancestors' quest to live in harmony with nature, because dualism—including the balance between humans and nature, humans and gods, and men and women—is commonly reflected in cosmogonic genealogies.

These historical accounts are foundational to a Kanaka geography and a sense of place because they identify and connect Kānaka to our kūpuna (ancestors; elders) and our ancestral homelands. A unifying theme that runs through these histories is the importance of knowing one's roots. As our kūpuna remind us, "I ulu nō ka lālā i ke kumu" (The branches grow because of the trunk/source) (ON 137). Indeed, a family thrives because of the roots and foundation laid by one's ancestors.

Chapter 2 Places to Stand

Ancestral Kānaka placed great importance on their personal space, as evidenced by the varying degrees of invisible boundaries that surrounded one's body. The level of intimacy and interaction that Kānaka enjoyed with each other in ancestral times was directly impacted by their societal status and genealogical lineage.[1] Wearing someone else's clothes, for instance, was largely forbidden.[2] High-ranking ali'i and their attendants went to great lengths to protect the personal effects of the ali'i.[3] Even amongst the maka'āinana, exchanging clothing was generally limited to those who were extremely close. The ancestral saying, "Na kahi ka malo, na kahi e hume" (The loincloth belongs to one, another wears it) suggests a very close blood relationship existed between two people because people generally did not share clothing.[4] Perhaps only someone who was related genealogically, of "ku'u ēwe, ku'u piko, ku'u iwi, ku'u koko" (my umbilical cord, my navel, my bones, my blood) (ON 207), or someone who was so closely related as to claim to know everything about her relative with the phrase "'Ike au i kona mau po'opo'o" (I know all of his/her nooks and crannies) (ON 131) might be allowed to wear a relative's personal effects. These clothing practices demonstrate how ancestral Kanaka customs were largely prescribed by one's status in society.

This chapter explores the construction of Kanaka identities, examining the social norms that prevented physical interaction between the ali'i and maka'āinana, which actually led to the distinct behavioral patterns characteristic of each group. These societal differences strongly influenced how ali'i and makaʻāinana related to the ʻāina and to others within their respective classes in ways that were dependent on their own roles in society.

Mele koʻihonua reveal that all Kānaka are genealogically interrelated, yet ancestral Kanaka society was divided into four distinct strata in ancestral times: aliʻi, kahuna, makaʻāinana, and kauā.[5] Kapu (taboos) regulated spatial distances between aliʻi, their personal attendants and kāhuna, the general population, and the kauā (lowest class of ancestral Kanaka society).[6] Those of the highest-ranking aliʻi and lowest-ranking kauā segments of the population were both untouchable, and members of each group were therefore known as "akua." For the aliʻi, akua made reference to their godlike nature. Aliʻi were sometimes referred to as the "kauā" of their people.[7] This was a humbling term, acknowledging that although aliʻi were high ranking, they were servants to their people. Conversely, kauā received the designation of akua because they were likened to despised ghosts.

Unlike the untouchable aliʻi and kauā, the makaʻāinana enjoyed much more freedom in their interactions with other makaʻāinana, embracing one another, exchanging hā (breath) by touching noses, and engaging in other forms of intimacy.[8] In spite of the personal boundaries that forbade intimate interaction amongst aliʻi and makaʻāinana, the two segments of the population shared great bonds of affection; the aliʻi cared for the people and the people honored the aliʻi by providing the aliʻi with harvests from the land and sea.[9]

By separating Kānaka into various classes, aliʻi distanced themselves from lower-ranking individuals, creating their own places in society. In order to maintain exclusivity, certain insignia and privileges reserved for royalty were required of aliʻi, including moʻokūʻauhau descending directly from the akua, mele inoa (name chants), hōʻailona (royal insignia), sacred birthplaces, ʻaha cords at the entrance of the home, wealth in the form of ʻāina and workforce, kapu status, birthright to perform ceremonies on heiau, and ultimately, sacred burial places.[10]

Verticality was of the utmost importance to aliʻi.[11] The higher the rank of an aliʻi, the closer to the akua they were. The strict kapu associated with aliʻi ensured exclusivity. The highest-ranking aliʻi were known as nīʻaupiʻo (arching of the coconut frond), which metaphorically implied that the aliʻi of this rank were born of the same frond, or family, because they were born of a successive and uninterrupted line of gods; not a single low-ranking aliʻi was among the kūpuna of a nīʻaupiʻo.[12] Although ʻĪʻī, Kamakau, Malo, and Pukui have varying descriptions for nīʻaupiʻo relationships, they all agree that the highest-ranking aliʻi nui were entitled to kapu reflecting their sanctity.[13]

Most of these scholars agree that within the nīʻaupiʻo rank there were at least two subcategories, "piʻo" and "naha." In order to maintain the status of their offspring, nīʻaupiʻo sometimes had children with their siblings.[14] When a full brother and sister pair born to nīʻaupiʻo parents had offspring, their children were known as "aliʻi piʻo," the highest nīʻaupiʻo rank possible.[15] These aliʻi were often carried from place to place by their attendants, as their very footsteps were considered sacred.[16] Aliʻi of this status often conversed with people and traversed at night, lest their shadows be cast on the ground, making it kapu. Even some aliʻi nui, outside of the nīʻaupiʻo piʻo rank, were required to remove their clothing and prostrate in the presence of these aliʻi, whose kapu were so great that they were considered to be like fire and blazing heat.[17]

The union between a half brother and half sister of nīʻaupiʻo parents produced children of the naha (curved) caliber, a rank that by some accounts later became known as hoʻi.[18] They were entitled to the kapu noho (crouching taboo).[19] Even children resulting from a father-daughter or uncle-niece relationship were considered to be of naha rank, provided that both of their parents were nīʻaupiʻo.[20] While naha were high-ranking aliʻi, they did not outrank aliʻi piʻo.[21]

Below the nīʻaupiʻo were the aliʻi wohi.[22] Like their nīʻaupiʻo counterparts, aliʻi wohi were aliʻi nui. They were born of a nīʻaupiʻo parent along with a parent of some other aliʻi nui status.[23] While other aliʻi nui who were not of nīʻaupiʻo status participated in the kapu moe (prostrating taboo), the aliʻi wohi were exempt. It was their responsibility to ensure that proper ceremonies were performed to honor the nīʻaupiʻo. They were free to move about in the presence of an aliʻi nīʻaupiʻo without fear of being sentenced to death for not prostrating.[24]

	Rank	*Relationship of parents*	*Kapu*
Aliʻi Nui	Nīʻaupiʻo: piʻo	Full siblings of nīʻaupiʻo parents	Kapu moe
	Nīʻaupiʻo: naha (hoʻi)	Half siblings of nīʻaupiʻo parents; parent-child of nīʻaupiʻo parents; uncle-niece/aunty-nephew of nīʻaupiʻo parents	Kapu noho
	Aliʻi wohi	Nīʻaupiʻo parent with a parent of some other aliʻi nui status	

Below the ali'i nui were various levels of kaukau ali'i. These lesser-ranking chiefs, while still of ali'i genealogy, did not possess the same kapu as their ali'i nui counterparts. Nevertheless, they too strategically planned who they had children with in order to elevate the status of their offspring, ensuring them a place of reverence in society. By producing descendants with partners who had moʻokūʻauhau that equaled or exceeded that of their own, ali'i were able to secure mana for their offspring and ensure that their own genetic legacy endured.[25]

Moʻokūʻauhau were venues by which ali'i were able to gain mana, and in ancestral times, before 'ōlelo Hawai'i (Hawaiian language) was written, the moʻokūʻauhau of ali'i were memorized and chanted at sacred ceremonies to acknowledge the sanctity of ali'i. Committing moʻokūʻauhau to memory linked descendants to their kūpuna and memorialized the histories and accomplishments of their kūpuna. The higher the rank of their kūpuna, the higher their own rank and the closer they were to the akua in the heavens.[26]

Genealogists backed their ali'i by proving that the ali'i they represented were of an uncontested high-ranking moʻokūʻauhau. Likewise, it was the responsibility of these genealogists to find flaws in the moʻokūʻauhau of their rivals. Disputing the moʻokūʻauhau of other ali'i could have the desired effect of reducing another person's mana, ties to the 'āina, and right to rule.[27] A person of uncertain parentage, for example, was referred to as "he kanaka no ka malu kukui" (a person from the shade of the kukui tree) (ON 75). Likewise, it was said of a person whose paternity was suspect that "komo wai 'ē 'ia" (another liquid had entered) (ON 198). Such a child might also be termed "he keiki kāmeha'i" (a wondorous child), for his paternal ancestry was undisclosed (ON 76).

One's moʻokūʻauhau greatly impacted one's rank in society. The 'ōlelo no'eau, "he ali'i nō mai ka pa'a a ke ali'i; he kanaka nō mai ka pa'a a ke kanaka" (a chief from the class of chiefs; a commoner from the class of commoners) reminds us that a chief is a chief because his ancestors were; a commoner is a commoner because his ancestors were. This statement warned those of royal descent to ensure a high-ranking lineage for their descendants by having children with other high-ranking ali'i (ON 63). Figure 1 illustrates how Maui ali'i strategically produced offspring with ali'i nui of other islands to establish senior lines on more than one island, securing political alliances for themselves during their own reigns. This is exemplified by the moʻokūʻauhau of Ka'ulahea I through Kahekili (also known as Kahekili II), the last ruling mō'ī (monarch) of Maui.

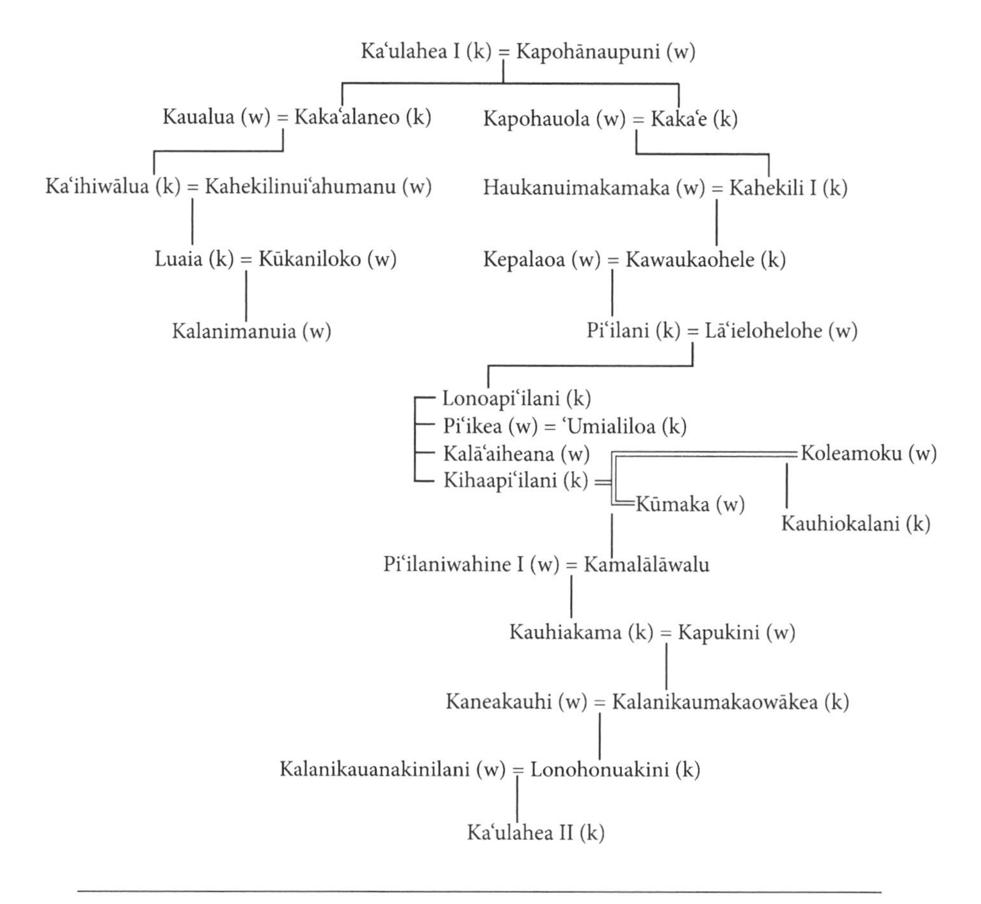

FIG. 2.1. Moʻokūʻauhau of select Maui aliʻi

SOURCES: Kamakau, *Ka Nupepa Kuokoa*, September 30, 1865, and *Ka Nupepa Kuokoa*, July 26, 1901.

Kaʻulahea I was a famous aliʻi of Maui. He had two children with Kapohānaupuni[28] (also known as Kapohānauaupuni[29]), Kakaʻalaneo and Kakaʻe, both of whom became important aliʻi of Maui. Kakaʻalaneo had a child with Kanikaniʻāʻula,[30] the daughter of Kamauaua of Molokaʻi, who gave birth to a male child, Kaululāʻau.[31] Kaululāʻau was sent to the island of Lānaʻi where he was made mōʻī. Kakaʻalaneo also had a child with Kaualua. From this union, Kaʻihiwālua[32] was born. Kaʻihiwālua and Kahekilinuiʻahumanu were the parents of Luaia, who entered into a relationship with Kūkaniloko, a female mōʻī of the island of Oʻahu and the daughter of Piliwale, the mōʻī of Oʻahu before Kūkaniloko. Kūkaniloko's grandfather was Kalonaiki and was also an aliʻi of

Oʻahu. Together, Luaia and Kūkaniloko had Kalanimanuia, who was made mōʻī after the passing of her mother, Kūkaniloko.[33] With Kapohauola, Kakaʻe had Kahekili I.[34]

Makua kāne (father)	*Makuahine (mother)*	*Kama (child)*
Kaʻulahea I	Kapohānaupuni	Kakaʻalaneo
	(Kapohānauaupuni)	Kakaʻe
Kakaʻalaneo	Kanikaniʻāʻula	Kaululāʻau
Kakaʻalaneo	Kaualua	Kaʻihiwālua
Kaʻihiwālua	Kahekilinuiʻahumanu	Luaia
Luaia	Kūkaniloko	Kalanimanuia
Kakaʻe	Kapohauola	Kahekili I

The relationships of Kakaʻalaneo and Kanikaniʻāʻula of Molokaʻi and Luaia's relationship with Kūkaniloko[35] of Oʻahu reveal how these partnerships were often calculated, strategic moves on the part of the aliʻi. Female aliʻi of the highest ranks were often closely monitored to ensure their virginity so that the father of their first child was of undeniable paternity. The children of such high-ranking unions were able to claim the genealogies of more than one island and potentially reign as aliʻi on all of the islands of their genealogies. A child of such a union cemented the bonds between the father's and mother's families.[36] The phrase "ua puka a maka" (the face emerged and is seen) was applied to the offspring of such a union because, throughout the life span of the child, the paternal and maternal families acknowledged their kinship with one another (ON 312).

An important Maui moʻokūʻauhau is that of Kahekili I, who had a relationship with Haukanuimakamaka.[37] Together they had Kawaokaohele (also known as Kawaukaohele),[38] a male child, who in turn had a child with Kepalaoa.[39] She gave birth to Piʻilani, a famous aliʻi of the island of Maui.[40] Piʻilani's mate, Lāʻielohelohe, was the daughter of Keleanuinohoʻanaʻapiʻapi (sister of Piʻilani's father, Kawaokaohele) and Kalamakua, who descended from the Oʻahu line through Kalonanui.[41] This nīʻaupiʻo pairing of Piʻilani and Lāʻielohelohe produced akua by the names of Lonoapiʻilani, Piʻikea, and Kihaapiʻilani.

Makua kāne (father)	*Makuahine (mother)*	*Kama (child)*
Kahekili I	Haukanuimakamaka	Kawaokaohele
Kawaokaohele	Kepalaoa	Pi'ilani
Pi'ilani	Lā'ielohelohe	Lonoapi'ilani
		Pi'ikea
		Kihaapi'ilani

Lonoapi'ilani became mō'ī after his father; however, when his brother Kihaapi'ilani started making lo'i (wetland taro gardens) to feed Kihaapi'ilani's supporters, Lonoapi'ilani saw this as a sign that his younger brother was rebelling against him. The two fought several times; each time Lonoapi'ilani was the victor. However, after Kihaapi'ilani traveled to the island of Hawai'i to seek the assistance of 'Umialīloa, his sister Pi'ikea's companion,[42] Kihaapi'ilani was finally able to defeat Lonoapi'ilani and become mō'ī of Maui.[43]

Kihaapi'ilani had a child with Kūmaka of Hāna: the famous ali'i, Kamalālāwalu.[44] Kamalālāwalu had a child with his niece, Pi'ilaniwahine I, from the Lonoapi'ilani line. That child was Kauhiakama, who had a relationship with Kapukini.[45] They had a child, Kalanikaumakaowākea, who paired with Kaneakauhi.[46] This union produced Lonohonuakini, who then partnered with Kalanikauanakinilani to beget Ka'ulahea II, a mō'ī of Maui.[47] Kekaulike was born to Ka'ulahea II and Papaikanī'au.[48] By having children with Papaikanī'au, Ka'ulahea II became the kupuna of many Maui ali'i. Ka'ulahea II also had a child with Kalanikauleleaiwi: Keku'iapoiwa Nui I.[49] Together, Kekaulike and Keku'iapoiwa Nui I had three famous ali'i of Maui: Kamehamehanui, Kalola, and Kahekili.[50]

Makua kāne (father)	*Makuahine (mother)*	*Kama (child)*
Kihaapi'ilani	Kūmaka	Kamalālāwalu
Kamalālāwalu	Pi'ilaniwahine I	Kauhiakama
Kauhiakama	Kapukini	Kalanikaumakaowākea
Kalanikaumakaowākea	Kaneakauhi	Lonohonuakini
Lonohonuakini	Kalanikauanakinilani	Ka'ulahea II
Ka'ulahea II	Papaikanī'au	Kekaulike
Ka'ulahea II	Kalanikauleleaiwi	Keku'iapoiwa Nui I
Kekaulike	Keku'iapoiwa Nui I	Kamehamehanui
		Kalola
		Kahekili II

Kalola (Kahekili's full sister), Kekaulike's daughter by Keku'iapoiwa Nui I entered into a relationship with a Hawai'i Island ali'i nui, Kalani'ōpu'u, and together they had Kiwala'ō.[51] Some genealogists suggest Kahekili was the biological father of Kamehameha. Kamehameha himself was unaware of this possibility until he was an adult.[52] If it is true that Kahekili was his father, there would have been no need for the islands of Hawai'i and Maui to battle and for Kamehameha to defeat his own half brother, Kalanikūpule, because Kamehameha could have ruled on both islands as his birthright.[53] It is interesting to note that Keku'iapoiwa Nui I gave birth to a famous Maui ali'i by the name of Kamehamehanui. Keku'iapoiwanui of Hawai'i Island later gave birth to a famous ali'i by the name of Kamehameha (also known as Kamehameha Nui [Kamehameha the Great] and Kamehameha I).

Another important mo'okū'auhau to discuss here is that of Kahahana, ruler of the islands of O'ahu and Moloka'i.[54] Kahahana was the great-grandson of Kalani'ōmaiheuila, grandson of Kalanikahimakeiali'i, and son of Ka'ionuilalaha'i, sister of Peleiōhōlani (mō'ī of O'ahu), and cousin of Kahekili (mō'ī of Maui). With 'Ēlani, an ali'i of 'Ewa, O'ahu, Ka'ionuilalaha'i had Kahahana.[55] In spite of the fact that Kahekili raised Kahahana as his own hānai (adopted) son, he eventually had Kahahana slain so that he alone could rule O'ahu.[56]

While Kahekili prayed to his war akua, Kūkeolo'ewa, for assistance in battle, war was not the only way to gain control of other islands; as very intelligent strategists, ali'i were also able to control other islands through marriage.[57] As mentioned previously, the union of Pi'ikea of Maui and 'Umialīloa of Hawai'i Island is a prime example, as are Kekaulike's five relationships with Kahawalu of Hāna, Keku'iapoiwa Nui I of Maui, Kanealae of Moloka'i, Hōlau of Kaua'i, and Ha'alo'u of Kaua'i, O'ahu, and Hawai'i.[58] Kekaulike's descendants became powerful ali'i nui on all of these islands, and thus they determined both control over great areas of 'āina and the course of our history.

Ali'i relied on their mo'okū'auhau to ensure their right to rule.[59] They also sought to prove that their mo'okū'auhau descended directly from the akua, as was the case of the chiefly families that descended directly from the god Lonoikamakahiki, or "nā lālā kapu a Lono" (the sacred branches/descendants of Lono) (ON 246), and Kalākaua, who used the *Kumulipo* to trace his genealogical line back to the creation of the world and the akua.[60]

Those who were not of such high lineage, especially those of junior lines, were encouraged to "kūneki nā kū'auhau li'ili'i, noho mai i lalo; ho'okahi nō, 'o ko ke ali'i ke pi'i i ka 'i'o" (let the lesser genealogies flow away and remain below; let that of the chief ascend), or to elevate the genealogy of the chief, not of one's own, lower-ranking genealogy (ON 206). Likewise, the junior lines were reminded to respect the senior line in the phrase, "i pa'a i kona kūpuna 'a'ole kākou e puka" (if our maternal ancestors died in bearing our grandparents, we would not have been born) (ON 136).

Birth order also affected an ali'i's status. For example, there was contention after the death of Kekaulike because his son Kamehamehanui was made mō'ī despite being a younger sibling to Kauhi'aimokuakama. As the elder son of Kekaulike, Kauhi'aimokuakama felt he should have been made mō'ī. Kamehamehanui was selected instead, based on the nī'aupi'o relationship of his parents, Kekaulike and Keku'iapoiwa Nui I; Kamehamehanui's lineage was therefore the line of akua.[61]

For those of lower-ranking mo'okū'auhau, increasing the genealogical mana of their descendants was difficult, as people with high-ranking genealogies typically preferred relationships with those of the highest mo'okū'auhau. A person with low mo'okū'auhau was referred to as "he i'a mo'a 'ole i kālua" (an uncooked fish), because there was nothing that a low-ranking person could do to increase his or her genealogical rank (ON 70). The kauā, though, had the lowest ranking genealogy in society. It was considered "hūpōenui" (extremely stupid) for a non-kauā to mingle their blood with a kauā (ON 123).

Rank and genealogy did not only determine one's status in society, but it also often impacted one's mobility. Although maka'āinana were free to move if they so desired, as their name suggests, they lived "on the land." Maka'āinana often remained on their ancestral lands generation after generation regardless of changes in government.[62] In contrast, ali'i did not usually enjoy the same rights of long-term residency to which the maka'āinana were accustomed, as ali'i land stewardships were redefined by wars and alliances.[63] As a result of the temporality sometimes associated with chiefly 'āina, ali'i either constructed ties to newly conquered land or reconstructed ties to reappropriated places. In cases where ali'i maintained rule for generations, such ali'i were celebrated as "mauliauhonua," meaning "descendant of old chiefs of the land," because such royal families were firmly established in their homeland.[64]

While many 'ōlelo Hawai'i words connect the maka'āinana and the 'āina, the terms that referred to the ali'i actually elevated their status by drawing a connection between the ali'i and the heavens. As akua that lived on earth, ali'i were often referred to as "Kalani" (the heavenly one), drawing a connection to the kūkulu o ka lani (borders of the sky).[65] Even the stars themselves were considered to be extensions of the highest-ranking ali'i, with some serving as 'aumākua (family deities) for the ali'i.[66] Stars were perhaps the most distinguished 'aumākua possible because of their positions in the highest stratum of the sky. More than any other known 'aumākua, the stars of the heavens had the best vantage point to watch over those that honored them. Stars shone over the ali'i, guiding them to the pono (virtuous) path.

Even cloud formations and other hō'ailona (divine omens or signs) that appeared in the heavens were connected to various ali'i.[67] The sight of a rainbow in the sky was a mark of royalty, especially for the nī'aupi'o and pi'o (arching classes of ali'i). Rain and rainbows were considered to be traces of chiefly footsteps in the 'ōlelo no'eau, "he 'ehu wāwae no kalani" (ON 65). Births and deaths of many ali'i nui were marked by strong storms that brought thunder, lightning, rain, and flooding. The 'ōlelo no'eau, "he kapu nā pōhaku hānau ali'i" (thunder during a chief's birth [is a sign] of sanctity) acknowledged the correlation between the birth of ali'i and the appearance of storms (ON 75). In some cases, very unusual occurrences such as the running of enormous schools of the red 'āweoweo fish were noted at times of historical significance.[68] From royal signs that appeared in the heavens to natural phenomena of the land and sea, such hō'ailona symbolized the sacredness and mana of the ali'i, reaffirming the chief's right to rule over the 'āina. While all Kānaka could claim a genealogical connection to the 'āina, the ali'i had a special relationship that differed from that of the maka'āinana. While the maka'āinana lived on the land, the ali'i were the akua—manifestations of the heavens—who ruled the land.[69]

The mo'okū'auhau of high-ranking ali'i shaped their identity and status. Although the ali'i may have been akua that lived on earth, they were also the younger siblings of the 'āina; therefore, ali'i who were pono were mindful of their connection and responsibility to the 'āina. In comparison to the many terms referring to the connection between maka'āinana and the 'āina, few do the same for the ali'i and the 'āina. Many of the words that do exist demonstrate a greater connection of the ali'i to the akua and heavens than to the land. When ali'i were likened to the 'āina, the steepest, most inaccessible places were usually selected. The proverb "ka 'ulu loa'a 'ole i ka lou 'ia," for example, suggests that an ali'i of very high rank was like a breadfruit that could not be plucked even with the assistance of a picking pole, which illustrates the exclusivity that came with chiefly status (ON 176). Another famous saying, "he pali lele a koaʻe," compares a high chief to a cliff that is too steep to climb, as does the ʻōlelo noʻeau, "he pali mania nā liʻi" (the chiefs are sheer cliffs), meaning they are not easily accessible (ON 95). Aliʻi were even compared to kites soaring up in the sky until "moistened by cold raindrops" in the saying, "he lupe lele a pulu i ka ua ʻawa," implying that such a person had risen to high ranks (ON 85). Other ʻōlelo Hawaiʻi terms are used to express prominence: "keha" (height, high, prominent); "kiʻekiʻe" (height, tallness); "kūhaka" (of high position, as one possessed by the gods); "kūpapalani" (state of heavenly foundation); "lani" (heavenly one); "lani kua haʻa" (poetic name for a very high chief: the highest heaven); "lani nuʻu" (highest heaven); "leʻolani" (lofty, tall, of chiefly rank); and "mamao" (distant).

Indeed, some of the most common words that reveal a chiefly connection to the ʻāina are "ʻai" and "kū," both implying the sense of ruling over the land. The term "aliʻi ʻai moku," for example, referred to chiefs who ruled over a moku (a large land division or district).[70] ʻAi also means "to eat off the land,"[71] while kū implies being firmly planted and standing on the ʻāina.[72] As long as an aliʻi is a good ruler, the needs of everyone under his or her leadership are met.

The relationship between the aliʻi and ʻāina is apparent in ʻōlelo noʻeau. Inasmuch as aliʻi were synonymous with akua that lived on earth, the ʻāina itself was highly revered by the aliʻi. It is no wonder, then, that many Kānaka believed that "he aliʻi ka ʻāina, he kauā ke kanaka" (the land is a chief; the people are its servants) (ON 62). This proverb suggests that the ʻāina even outranked the highest of aliʻi. This ʻōlelo noʻeau gives us a glimpse into a worldview in which the aliʻi had the kuleana to serve the ʻāina by making it productive. Aliʻi fulfilled this obligation by creating innovative farming techniques such as ʻauwai (ditches) that fed into wetland loʻi that in turn emptied into loko iʻa (fishponds). In this manner, aliʻi who were pono were able to sustain a large population. Aliʻi who had vast landholdings and cared for the ʻāina and the makaʻāinana were often content. The ʻōlelo noʻeau "he māʻona moku" describes those people who were satisfied with their landholdings (ON 88). Like the term "ʻai," referring to aliʻi that ate off the ʻāina, "māʻona" was applied to those who had content stomachs, having had enough to eat.

Aliʻi respected their makaʻāinana because the makaʻāinana farmed the ʻāina and offered produce to the aliʻi. Therefore, aliʻi ensured the makaʻāinana had land to reside on but insisted that "no ka noho ʻāina ka ʻāina" (the land belongs to the one dwelling on the land) (ON 254). Land was granted to a family, but whoever chose to move away risked forfeiting their rights to the ʻāina; those who maintained their connection to the ʻāina were the stewards.

Maintaining a physical connection to one's kulāiwi (ancestral homeland; plains of ancestral bones) ensured the continuation of a family's legacy at a particular place. A person who honored his or her kulāiwi was compared to a warrior in the phrase, "he maka lehua no kona one hānau" (a warrior [with the] face of the sands of his/her birth), meaning such a person was loyal and honored in his or her birthplace (ON 86). This suggests that, like a warrior who remains loyal and steadfast to the aliʻi, people who dwell on their kulāiwi possess the same admirable traits. Loyalists were highly prized by the aliʻi because it was believed that "ʻaʻohe e nalo ka iwi o ke aliʻi ʻino, ʻo ko ke aliʻi maikaʻi ke nalo" (the bones of an evil chief will not be hidden, but the bones of a good chief will) (ON 17). Aliʻi depended on their people for sustenance while they were alive and to protect their bones when they died. By caring for the land and the people, aliʻi ensured that their bones would be well cared for after their deaths. While it is true that a particular ʻōlelo noʻeau stated, "ʻO ke aliʻi ka mea ikaika, ʻaʻole ʻo ke kanaka" (It is the chief who is strong, not the general population) (ON 267), without the backing of one's makaʻāinana, an aliʻi was powerless to accomplish great feats and ward off rival aliʻi. This saying suggests that by themselves, the general population was limited according to their individual abilities; however, aliʻi had the power to oversee large numbers of people to accomplish their goals.

Many poetic references to aliʻi and certain places were coined as well, reinforcing the ties of aliʻi to specific ʻāina. "Kalo kanu o ka ʻāina" (taro planted on the land) favorably compared people to the kalo of the ʻāina (ON 157). An aliʻi of this nature was a native to the area that he or she ruled, and like the kalo, such an aliʻi had permanent roots in the ʻāina.[73] "Nā Hono a Piʻilani" (the Bays of Piʻilani) (ON 243)[74] was an example of the permanence of these roots. It was a name applied to the island of Maui, acknowledging the powerful aliʻi nui, Piʻilani, and the six famous bays of West Maui: Honokahua, Honokeana, Honokōhau, Honokōwai, Honolua, and Hononana. These bays overlook the islands of Kahoʻolawe, Lānaʻi, and Molokaʻi, all of which were also under the rule of this famous Maui aliʻi.[75] Kamalālāwalu, another aliʻi nui of Maui, was honored by similar references: "Mauiakama" (Maui, the island of Kama[lālāwalu]) and "Mauinuiakama" (Maui and its surrounding islands of Kama[lālāwalu]).[76] Referring to an entire island by the name of an outstanding ali'i was perhaps the highest honor bestowed upon royalty, as that ruler's legacy was fixed in history to be remembered by future generations.

While poetic references to whole islands heralding the names of ali'i was a great honor for the ali'i for whom the islands were named, not all lands were of equal importance to the ali'i. Securing sacred lands, regardless of size, under one's rule was an avenue for ali'i to increase their own mana. Two Maui examples clearly articulate this point. In the first example, Kahekili, a very famous ali'i nui of Maui, raised his nephew Kahahana as his own adopted son, as noted earlier. Kahahana, because of his genealogical ties, later became the ali'i nui of the island of O'ahu. After some time, Kahekili approached Kahahana and requested that he be given the ahupua'a (land division) of Kualoa. Initially, Kahahana was desirous to comply with this request as a symbol of loyalty to his hānai father. However, he decided against such action after his kahuna warned him that Kualoa was more than just a piece of land; whoever ruled the sacred ahupua'a of Kualoa and the whale ivory that washed up on its shores would ultimately govern the entire island of O'ahu.[77] Kahekili was enraged at Kahahana's refusal and brutally ended his nephew's life, and as a result he took over O'ahu.[78]

In the second Maui example, already introduced, two sons of the famous ali'i nui Pi'ilani were at odds with one another. Pi'ilani bestowed upon Lonoapi'ilani, as the elder son, the birthright of ruling the island of Maui. One day, Lonoapi'ilani became outraged when he discovered that his brother, Kihaapi'ilani, was feeding the people of Maui. Lonoapi'ilani saw this as Kihaapi'ilani's attempt to usurp his elder brother's mana. Lonoapi'ilani therefore banished Kihaapi'ilani, forcing him into poverty and anonymity.[79]

Kihaapi'ilani entered into a relationship with a woman, Koleamoku, whose father was a strong supporter of Lonoapi'ilani. After the couple had a child, Koleamoku approached her father, Ho'olaemakua, and requested to be given 'āina for her family. Her father granted her the entire district of Hāna, but she declined, stating that her husband did not want the entire moku of Hāna; rather, a few choice lands would suffice. Upon revealing the particular 'āina Kolea's husband was requesting, Ho'olaemakua became enraged. He realized that only an ali'i would pass up the opportunity to receive the entire Hāna district in exchange for smaller parcels that contained the resources needed for battle. Knowing that only an ali'i who knew eastern Maui well and who wanted to overthrow Lonoapi'ilani would make such an appeal, Ho'olaemakua replied to his daughter, “He alii ko kane, aole he kanaka e ko kane, he alii ko kane o Kihaapiilani, a he alii ko keiki, aole au e make mahope o ko kane, aia no hoi mahope o ke kaikuaana ona e make ai keia mau iwi, aole i makemake ko kane i wahi mahiai no olua aka, i wahi e kipi ai i ke aupuni” (Your husband is no ordinary person. Your husband is the chief Kihaapi‘ilani. Your child is a chief. I shall not die for your husband. I shall remain steadfast behind his older brother till these bones perish. Your husband does not want farmlands for both of you; he seeks to rebel against the kingdom).[80]

As seen from these two examples, political maneuvering was often land based. Ali‘i gained mana and waiwai (wealth) through the ‘āina they ruled and would declare war in an effort to acquire new lands for such things as food production.[81] Whenever an ali‘i nui was successful in acquiring new ‘āina, one of his first obligations was to kālai‘āina (redistribute the land) amongst his supporters. Those who demonstrated the most allegiance to the ali‘i nui received the best ‘āina.[82]

While genealogy was important in determining a chief’s status in society, mo‘okū‘auhau by itself was not enough. In order to maintain one’s status and the respect of the maka‘āinana, an ali‘i had to acknowledge the akua as the source of one’s ‘āina and recognize the allies that had assisted along the way. In ancestral times, when new ali‘i nui ruled, they would usually perform a kālai‘āina. In this way, the ali‘i nui was able to reward supporters and maintain friendly relationships with these allies.[83] An ‘ōlelo no‘eau states, “he hānai ali‘i, he ‘ai ahupua‘a” (one who feeds a chief, rules an ahupua‘a), suggesting that those who care for a chief are often rewarded with large parcels of land (ON 66). Another ‘ōlelo no‘eau offers the same sentiment: “ke kaulana pa‘a ‘āina o nā ali‘i” (the famous landlords of the chiefs), meaning that the best warriors were awarded the best lands (ON 187).

Redistributing the ‘āina and defining parcels of ‘āina not only facilitated the construction of physical place on the landscape but, just as importantly, created corresponding places in society, niches from which people of varying ranks stood. For the ali‘i, land sections represented a division of power and control. A single ali‘i nui might rule over an entire mokupuni (island). Below him or her, ali‘i ‘ai moku were responsible for the welfare of the moku, its resources, and its people. Similarly, konohiki or ali‘i ‘ai ahupua‘a cared for the people within an ahupua‘a.

Unlike the other three major islands that were divided into five or six moku, the island of Maui was uniquely divided into twelve such moku.[84] Providing supporters of an ali'i with rights to rule over land was perhaps the best way to maintain friendly relations with allies. The fact that Maui, a double island, had twelve moku suggests a number of things. First, Maui must have been very rich in resources; only a fertile island could support so many moku. Second, by dividing the island in this manner, the ali'i nui had twelve moku to allot amongst supporters; this was twice as much as any other island. Third, should an ali'i 'ai moku choose to someday rebel against the highest ali'i nui, that chief would only control one-twelfth of the land. Strategically, this was a clever move on the part of the ali'i nui, who probably realized that it would be difficult for ali'i 'ai moku on Maui to stage an uprising.

Below the ali'i 'ai moku were the ali'i 'ai ahupua'a, also known as konohiki, who had the responsibility of maintaining a balance within the ahupua'a.[85] The konohiki even had the kuleana of redistributing resources when necessary to ensure that each tenant's needs were met.[86] Of all ali'i ranks, the konohiki had the closest relationship with the maka'āinana because they served as liaisons between higher ranking ali'i and the maka'āinana.[87]

Yet, ali'i of all ranks demonstrated their mana by coordinating the efforts of the maka'āinana to construct long irrigation ditches, large fishponds, and extensive heiau. Only ali'i who had garnered the respect and admiration of the general population had enough manpower to achieve such laborious tasks.[88] Such huge undertakings did not go unnoticed by the akua and other ali'i.

Construction of Place by Maka'āinana

Unlike the ali'i, who demanded exclusivity to maintain their mana, maka'āinana were very inclusive, often gathering in large groups on special occasions. Social interaction renewed bonds with the community on a daily basis through acts of generosity. Sharing and exchanging one's material wealth and resources from the land and sea were characteristic of the maka'āinana. Acts of kindness and interdependence allowed the maka'āinana to prosper and to establish their own niches in society. Thus, the general population thrived on horizontal relationships rather than verticality, valuing each person's contribution to society.[89]

As 'ōlelo Hawai'i and 'ōlelo Kahiki (Tahitian) are closely related members of the same Polynesian language group, it is no coincidence that the Tahitian word "mata'einaa" closely resembles the word "maka'āinana." While mata'einaa refers to an ahupua'a, maka'āinana refers to the people who were the backbone of the ahupua'a, suggesting that the people and the land division are synonymous.[90] Similarly, the indissoluble relationship of the maka'āinana and the 'āina is reflected in various 'ōlelo makuahine (mother tongue) terms. The word "kua'āina" refers today to people who live off the 'āina and who carry the burden of the land on their backs. This term means "back of the land," which suggests that the 'āina is supported by the maka'āinana.[91] "Kama'āina" (child of the land) and "kupa o ka 'āina" (native of the land) similarly refer to the ties between the people and the 'āina.[92] Another term, "ēwe," meaning "sprout, rootlet, lineage, kin," once again intertwines the concepts of genealogy, growing from the land, and establishing one's roots.[93] The foundation of the Kanaka society, therefore, is the 'āina. One who resides on the same 'āina for an extended period of time is called an "'āpa'a," and "pa'a" implies that one is steadfast and committed to one's place.[94] These terms reflect the deep connection of the people to their ancestral places.

Being rooted in place was very important to many maka'āinana. Kalo was the staple food in ancestral times, and it could be likened to the maka'āinana, as both the kānaka and the kalo depended on the 'āina for their nourishment and well-being. A rootless Kanaka—one whose connection to the 'āina had been severed—suffered great psychological, emotional, and physical losses. In order for the maka'āinana to flourish, remaining "rooted" in one place was key.[95] People were considered an offshoot of the kalo—the 'ohā, a sign of health and growth. The 'ohana (family), a term derived from the offspring of the kalo, was the cornerstone of maka'āinana life.[96] Inevitably, families living on kulāiwi in close proximity to one another intermarried and formed lasting bonds with one another; strong ties to their 'ohana and ancestral places further strengthened their ties to the 'āina. The 'ohana had the additional function of encouraging love and respect amongst family members, and it was not unusual for well-established families to be related to most if not all of the people in their ahupua'a. By continuing to work the 'āina and partaking of the kalo, the maka'āinana maintained their family roots and lifestyle.[97]

Because the 'āina was family to Kānaka, it was not a commodity to be bought or sold, and it is therefore not surprising that no 'ōlelo makuahine word existed in ancestral times for the concept of land ownership. Instead, the word "kuleana" (small parcels of land awarded to maka'āinana, so called because the maka'āinana had the "responsibility" to care for these lands in perpetuity)was adopted during the time of the Māhele (the land division of 1848) and referred to lands held by the maka'āinana. According to C. J. Lyons, "Kuleana means originally a property or business interest in anything. The common people were in former times assigned certain portions of the chief's lands, to occupy at the will of the chief. Generally speaking, there was a good degree of permanence in this occupancy, provided that service was duly paid to the superior."[98]

Living on the 'āina was a great kuleana. The kuleana of the maka'āinana was multilayered. Each piece of 'āina had its own place on the landscape, and the people who resided there likewise had a specific place in society.[99] Maka'āinana had a kuleana to other maka'āinana. Indeed, those living upland traded their produce with those near the ocean.[100] Maka'āinana also had a kuleana to honor their konohiki and ali'i of varying ranks. During times of war, the maka'āinana might also be called upon to feed the warriors of their ali'i. Maka'āinana even had a kuleana to the akua. Praying and offering one's first catch or harvest to the akua was daily practice. During the Makahiki season, maka'āinana paid tribute to the god Lono by offering produce from the 'āina as well as material goods created by hand.[101] In post-settlement times, maka'āinana were required to serve the konohiki by working on royal lo'i every Pō'alima (Friday); thus such lo'i became known as Pō'alima.[102]

The maka'āinana, as noted earlier, often remained on the same kulāiwi generation after generation, and the 'ohana enjoyed a somewhat permanent stewardship and occupancy of the 'āina.[103] All significant events of one's life—from the birth of a child to the death of a kupuna—occurred on the kuleana.[104]

'Ōlelo no'eau inform us about the link of the Kānaka to the 'āina. "Ke ēwe hānau o ka 'āina" (the lineal descendants born of the land) is used in reference to Kānaka, who, like their kūpuna, are also born of the 'āina (ON 182). This wise saying associates Kānaka with the sprouts, rootlets, and kin of the 'āina, thereby reinforcing the genealogical connection between the two.

To have one's own place on the landscape ensured one's security and grounded a person. It was important for maka'āinana to have a place to call home and to maintain one's family's roots. A person who had these things was referred to poetically as "he lani i luna, he honua i lalo." Simply stated, the phrase translates, "a heaven above, an earth below"; in other words, a person who had a piece of 'āina to care for was said to have the sky above and the earth beneath (ON 79), which signifies everything one needs. Only a people so closely connected to the ʻāina would equate stewardship of ʻāina as the ultimate mark of security and stability.

One reason why the land was a mark of security and stability is because it was a final resting place for the dead. Makaʻāinana often buried their dead near their homes. "Kulāiwi" was not merely an abstract term that linked the living to the dead, but it was in reality both the homeland of the living and the burial ground of the dead. In contrast to the more open nature of makaʻāinana death practices, aliʻi burials were marked by great secrecy. Family members and/or attendants of high-ranking aliʻi would go to great lengths—securing aliʻi in caves and other remote locations—to ensure that aliʻi grave sites would not be found by grave robbers, who were interested in obtaining mana from the royal dead through their human remains and funerary objects.[105] While both the aliʻi and makaʻāinana may have been mortified by the thought of their bones being disinterred and disturbed, the locations of makaʻāinana burial sites were often well known by their families. As the mamo (descendants) of ancestors of a particular kulāiwi, succeeding generations had an obligation to their kūpuna to protect their bones and to care for their grave sites.[106]

In exchange, the mamo earned the spiritual guidance of all of the kūpuna that had gone before them. Even today, moʻopuna (grandchildren) are linked to one's kūpuna, both seen and unseen, for the ancestors continue to live through their descendants. In this way, the moʻo (successive line) continues generation after generation. Because of the connection between the ʻāina and the Kānaka and also between the Kānaka and their kūpuna, it makes sense that moʻokūʻauhau are linked to place. When reciting one's pedigree, it is important to acknowledge the one hānau (sands of birth) as well as one's kūpuna. Being able to map one's place on the landscape firmly rooted the makaʻāinana back to the places of their kūpuna. A person saying, for example, that they were from Kuewa implied that they and their family had roots on the island of Maui, in the moku[107] of Kāʻanapali, in the kalana (a division of land smaller than a moku) of Kahakuloa, on the ʻili ʻāina (a smaller land division than ahupuaʻa that was subdivided into moʻo ʻāina; also known as ʻili ʻāina) of Kuewa, a place ma uka (toward the mountains), bordering a stream. As a person residing on an ʻili ʻāina known to have twenty loʻi kalo, they were probably farmers (as were their kūpuna).

Kānaka today are often very proud of their ancestry, spending countless hours researching the names of their ancestors and the places that their ancestors call(ed) home. In spite of the fact that extensive knowledge of mo'okū'auhau was at one time reserved primarily for ali'i, the trend today is for Kānaka of all social backgrounds to research their own genealogies. This desire to know one's roots has been exacerbated by detrimental practices that have severed the bonds between the people and the 'āina. The Māhele, for example, paved the way for land privatization and ultimately disenfranchised the vast majority of Kānaka from their kulāiwi. As Kanaka attorney Melody MacKenzie asserts,

> The final step in the *Mahele* process was determining the interests of the *maka'āinana*. . . . While *kuleana* were generally among the richest and most fertile in the islands, there were a number of restrictions placed on *kuleana* claims. First, *kuleana* could only include the land which a tenant had actually cultivated plus a houselot of not more than a quarter acre. Second, the native tenant was required to pay for a survey of the lands as well as bring two witnesses to testify to the tenant's right to the land. It is estimated that of the 8,205 awards given by the Land Commission, 7,500 awards involved *kuleana* lands. This resulted, however, in only 26 percent of the adult male native population receiving such lands. The plan adopted by the king and chiefs for division of the land had stated that the *maka'āinana* were to receive, after the king partitioned out his personal lands, one-third of the land of Hawai'i. However, only 28,600 acres, much less than one percent of the total land, went to the *maka'āinana*.[108]

Similarly, exorbitant land tax rates, sometimes amounting to tens of thousands of dollars annually, have forced many Kānaka to sell their kulāiwi, as was the case with my own 'ohana. Many Kānaka are still suffering from the psychological effects of being uprooted from and deprived of our kulāiwi.

Nevertheless, the ancestral connection that families have with the 'āina continues to link people genealogically to one another and to their kulāiwi. Because Kānaka lived on their kulāiwi for many generations, modern Kānaka may access land records, documenting the names of those awarded the kuleana of caring for the 'āina, and for families that have kept their kulāiwi within the family, these land records reveal the names of each succeeding generation as they too took on the kuleana as land stewards. The 'āina, then, serves as a genealogical history of past generations. In this way, land records and mo'okū'auhau serve as mechanisms for readjusting life to reconnect with their kūpuna and their kulāiwi.

Unlike ali'i who needed to prove the sanctity of their genealogical lineages to maintain their birthrights to rule, maka'āinana did not need to reveal their mo'okū'auhau to maintain their status. Therefore, they were often careful not to disclose their mo'okū'auhau to others. The saying "mai kaula'i wale i ka iwi o nā kūpuna" warns people not to openly discuss one's mo'okū'auhau and expose the bones of one's kūpuna (ON 225). Moreover, the phrase "aia a pa'i 'ia ka maka, ha'i 'ia kupuna nāna 'oe" (only when your face is slapped, then you should reveal who your ancestors are) reveals that Kānaka were taught not to boast of their kūpuna; however, should a kupuna be insulted, it would then be appropriate to set the record straight (ON 6). Similarly, if someone claimed to have knowledge of one's mo'okū'auhau that was incorrect, an appropriate response might be, "'a'ohe 'oe no ko'u hālau" (you are not of my house) (ON 24). Asserting that a person was not related and therefore unacquainted with your mo'okū'auhau turned the tables and revealed that person's ignorance.

The ali'i and maka'āinana each had their own separate and distinct places and kuleana in society. Yet, in spite of the social boundaries that prevented their physical interaction with one another, each was reliant on the other. This sense of interdependence is also apparent in several 'ōlelo no'eau. "I 'āina nō ka 'āina i ke ali'i, a i waiwai nō ka 'āina i ke kanaka" means "the land is the land because of the chiefs, and the land is prosperous because of the general population" (ON 125). In other words, both the ali'i and maka'āinana have important roles to play in society: the ali'i oversee the 'āina, but the general population needs to work the 'āina in order to make it productive. Similarly, ali'i are reminded that "i ali'i nō ke ali'i i ke kanaka" (a chief is a chief because of the people who serve him or her) (ON 125). No matter how high ranking an ali'i might be, chiefs must not forget that they are nothing without the people who serve them and work the 'āina. Moreover, the saying "hānau ka 'āina, hānau ke ali'i, hānau ke kanaka" (the land is born, the chiefs are born, the general population is born) cautions ali'i that in spite of the varying ranks of people in society, the land, chiefs, and general population are all of one stock, sharing the same cosmogonic genealogy (ON 56).

Conclusion

From the highest-ranking chiefs to the general population, ancestral Kānaka had an undeniable connection to the 'āina. The social norms that prevented physical interaction between the ali'i and maka'āinana led to distinct behavioral patterns characteristic of each segment of the population. Ali'i had kapu that created barriers between themselves and lower-ranking people in society, and they had the kuleana of ensuring that the maka'āinana and 'āina were productive so that the needs of the people and the 'āina were in balance. While ali'i had the advantage of being able to acquire more 'āina through warfare, thus extending their domain and rewarding supporters with interests in their newly attained 'āina, it was equally possible for ali'i to be forced to forfeit lands should they be defeated in battle. As a result, ali'i were a highly mobile societal unit and were often forced to (re)construct ties to the places that they ruled.

Maka'āinana, on the other hand, generally enjoyed long-standing ancestral ties to the lands they lived on. In spite of wars and changes in government, the maka'āinana rarely feared being displaced from their kulāiwi. Their presence on the landscape was more permanent than the temporal rule of the ali'i. Maka'āinana had the kuleana of working the 'āina, fishing the sea, and honoring the ali'i. For these islanders, living in accord with nature and one another was key to their survival. Therefore, ancestral Kānaka maximized their resources and manpower by specializing in specific skills and becoming masters of their trades. Maka'āinana enjoyed close interaction with other maka'āinana. Extending over many generations, these relationships between families residing in the same ahupua'a were strong. Interdependence and reliance on one another created a societal bond that ensured that people of all ranks worked together for the common good of all.

Chapter 3 Fluidity of Place

Ka pae ʻāina Hawaiʻi is 1,200 miles north of the Pikoowākea, also known as ke alanui polohiwa a Kāne (the equator).[1] Ka pae ʻāina Hawaiʻi is sometimes referred to as Hoakalani (also known as Hoakaailani) because of its resemblance to the moon phase by the same name.[2] Others use the name Kūkūau, insisting that ka pae moku bears remarkable similarities to the shape of the kūkūau crab.[3]

The resemblance of ka pae ʻāina Hawaiʻi to the Hoakalani moon phase

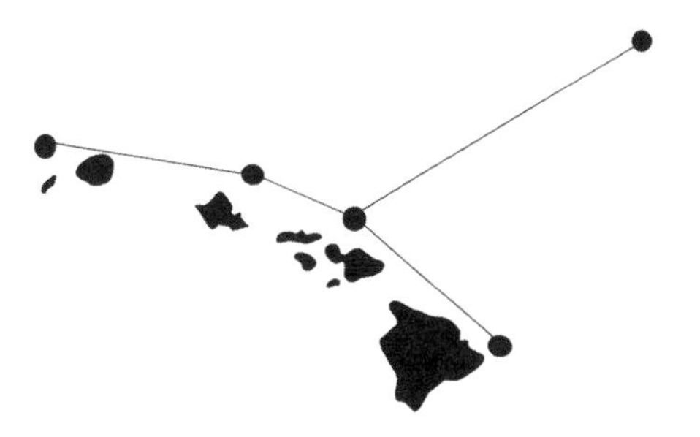

The resemblance of ka pae ʻāina Hawaiʻi to the Kūkūau constellation

Although many people are most familiar with the eight major islands, Kanaka scholars ancestrally acknowledged at least twelve islands: Hawaiʻi, Maui, Kahoʻolawe, Lānaʻi, Molokini, Molokaʻi, Oʻahu, Kauaʻi, Niʻihau, Lehua, Kaʻula, and Nihoa.[4] Over time, the names of some of these islands have changed. Kauaʻi was once known as Kamāwaelualani[5] and Kamāwaelualanimoku,[6] while Oʻahu previously went by the names Laloohoaniani, Lalowaia, Loloimehani, Oʻahualua, and Oʻahualuanuʻu.[7] ʻIhikapalaumaēwa, ʻIhikapulaumaēwa, Kūlua, and Mauiloa were older names for the island of Maui.[8] Hawaiʻinuiākea, Hawaiʻinuikuauli, Hawaiʻinuikuaulikaioʻo, Lononuiākea, and Nononuiākea were ancient names for Hawaiʻi Island.[9] The Hoakaailani tradition notes that Hawaiʻi was once known as Kuahiwi Kuauli (dark mountains), Maui as Hīhīmanu (ray fish), Lānaʻi and Kahoʻolawe as Huʻahuʻakai (sponges; sea foam), Molokaʻi as ʻAlaehuapī (tricky mudhen), Oʻahu as Ahumaunakilo (observatory mountain of Ahu), and Kauaʻi as Laukīʻeleʻula (dried red ti leaf).[10] According to Kamakau, the islands of Kauaʻi, Oʻahu, Maui,[11] and Hawaiʻi were named after the children of Papa and Wākea.[12] According to Johnson, Molokini was once known as Mololani. Molokaʻi was known as Molokaʻinuiahina. Lānaʻi previously went by the names Lānaʻiakāula, Lānaʻikāula, Lānaʻikāulawahine, Lanikāulawahine, and Nanaʻi. Kahoʻolawe was known as Kanaloa, Kahoʻolewa, Kohemālamalama, and Kohemālamalamaokanaloa. Mokupapapa was once known as Napapakahakuākeaolono.[13]

In addition to naming islands, Kānaka demonstrated a firm understanding of the depth of place, mapping their places by attaching names to the various regions of their environment. Their worldview was the product of the "world (they) view(ed)." This chapter demonstrates that their places were not confined to the boundaries on ʻāina but extended vertically and horizontally in every direction, encompassing heavenscapes, landscapes, and oceanscapes. By identifying and, more importantly, naming the various strata of the heavens, regions on the landscape, and depths of the ocean, Kānaka transformed spaces into personalized places.[14]

Heavenscapes

ʻŌlelo Hawaiʻi terminology reveals the degree to which ancestral Kānaka understood the complexity of their environment. In the case of the heavens, intricate layers of the lani (skies) were skillfully separated into distinct strata. The lani were referred to as "ka paʻa i luna" (the firmament above), while the region between the lani and ʻāina was known as "ka lewa."[15] The firmament attached to the ʻāina was "ka paʻa i lalo" (the solid below). The highest stratum was known as "lewa lani"; below the lewa lani was "kamakuʻialewa" (the circle of air surrounding the atmosphere). Following kamakuʻialewa was "keapoalewa" (the circle of space). Birds flew below, in the layer known as "lewa nuʻu." The area in which a person's feet would dangle when holding onto a tree with one's hands was known as "hakaalewa" (ladder space). Below this was the lewa hoʻomakua (the space created when one stands on one foot and raises the other foot).[16]

The saying, "Aia i ka ʻōpua ke ola: he ola nui, he ola laulā, he ola hohonu, he ola kiʻekiʻe" (Life is in the clouds: great life, broad life, deep life, elevated life), sheds some light on the importance of the heavens to Kānaka.[17] The proverb contends that someone who understands the omens of the heavens can ascertain from the clouds what the future holds in terms of rain, prosperity, or natural disaster. By utilizing the signs from the heavens, Kānaka were also able to determine the best times to fish, plant, and hunt, and navigators were able to detect remote islands in the distance.[18]

Inasmuch as the skies provided Kānaka with a connection to physical elements in the environment, they also bestowed a spiritual relationship with the akua. For kāhuna and aliʻi, the heavens were crucial to their places in society. It was of immense importance to the aliʻi and the kāhuna who served them to name and claim the heavens, thereby constructing a virtual place, far removed from the rest of the population for high-ranking aliʻi to figuratively reside. Like the heavens, the aliʻi—the physical embodiment of the akua on earth—were untouchable. Being able to make distinctions between layers of

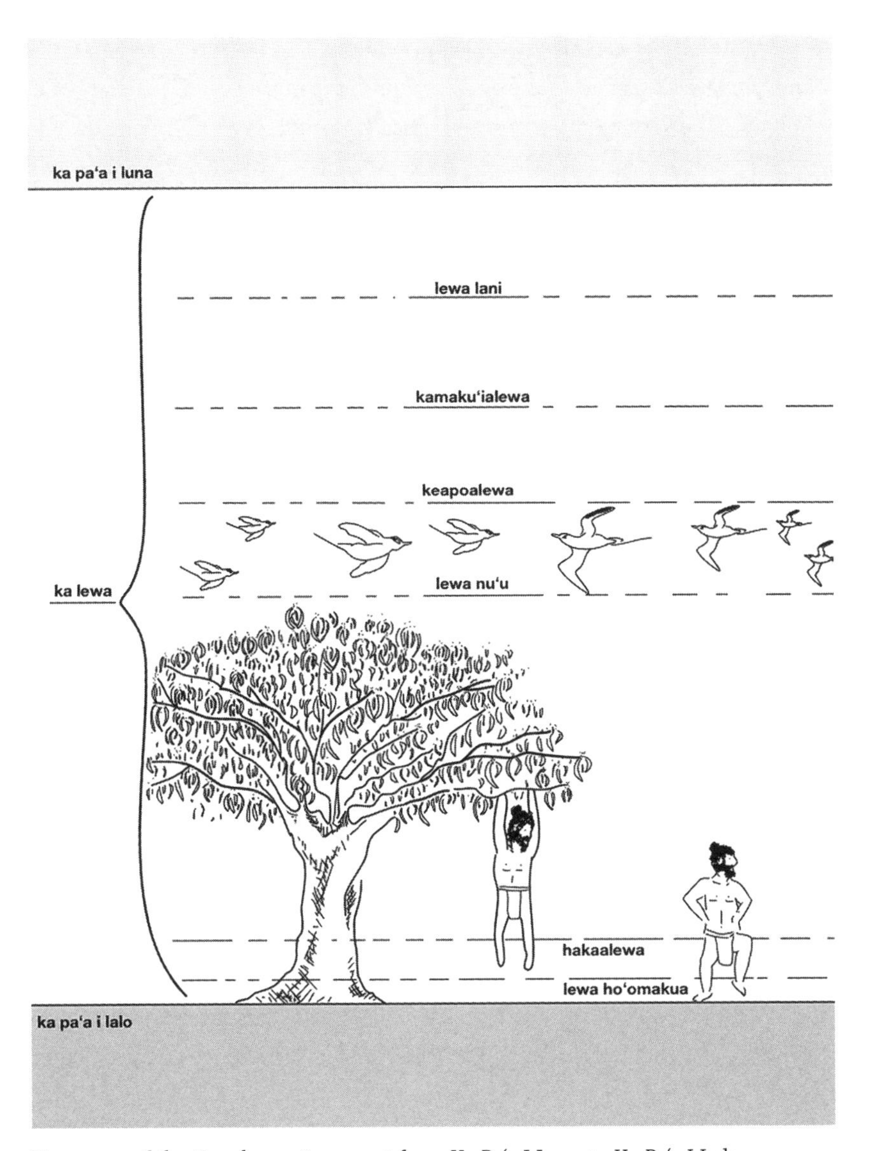

The strata of the Kanaka environment from Ka Paʻa I Luna to Ka Paʻa I Lalo

heavens, indistinguishable to others, provided mana to those who knew kilokilo (stargazing and the reading of omens). It is no wonder then that great scholars of the kahuna order devoted much time and effort into understanding the heavens. It was a kind of mystical power limited to a few; kilokilo was an art cloaked in deep secrecy. Through kilokilo, kāhuna were able to predict the future and detect omens in the sky, providing ali'i with an additional tool for decision making, influenced by the akua themselves.[19]

Failure to heed the advice of such kāhuna often led to the demise of disobedient ali'i. For instance, when Kamalālāwalu, ali'i of Maui, announced that he intended to war with the chiefs of Kohala, Kona, and Ka'ū, his prophet Lanikāula discouraged him, insisting that he and his supporters would be routed by the Hawai'i forces.[20] Outraged by the sentiment of his kahuna, Kamalālāwalu vowed to return victorious and burn Lanikāula alive for his inaccurate prophecy.[21]

Kamalālāwalu sent a swift runner to Hawai'i to determine the size of the population on the island of Hawai'i. The runner returned and reported that the Kona-Kohala Coast was not as heavily populated as previously estimated. Lanikāula, adamantly disagreeing, stated that the runner had not passed through some of the more densely populated areas. He also insisted that many people were at other locations—either upland or out to sea—at the time, and the runner's population figures were grossly underestimated.

Kamalālāwalu then went on to disagree with Lanikāula as to where the battles should take place. Rather than heeding the advice of Lanikāula—who was steadfast in his conviction that the battle be fought at Pōhakuloa, a site that would give the Maui forces the upper hand—Kamalālāwalu instead took the advice of Kauhipa'ewa and Kīhāpai'ewa, whom Kamalālāwalu would learn later were actually allies of the Hawai'i forces. As Lanikāula predicted, Kamalālāwalu and his warriors were slaughtered.[22] The hō'ailona Lanikāula relied upon had been accurate.

Landscapes

From the summits of the highest mountains to the lava flowing into the sea, the various parts of the landscape were likewise identified and named. Mauna, the highest places on the landscape, were covered with mist and fog. Kuahiwi were below the summit but above the tree line.[23] Within the mountainous regions of the mauna and kuahiwi were the kuahea,[24] wao nahele,[25]

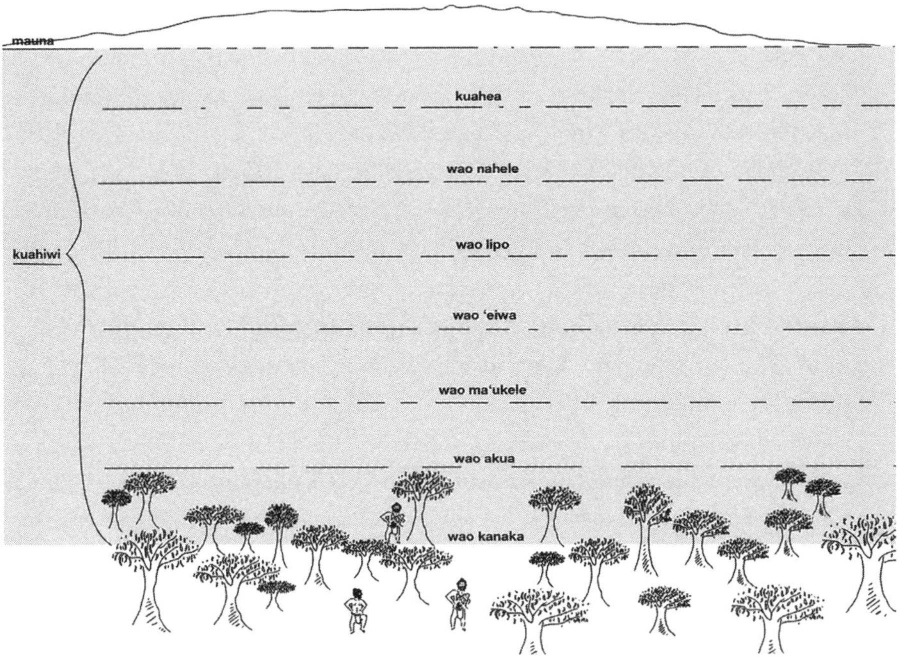

The regions of the mauna and kuahiwi

wao lipo,[26] wao ‘eiwa,[27] wao ma‘ukele,[28] wao akua,[29] and wao kanaka. Humans inhabited and cultivated this last zone, wao kanaka.[30]

Islands themselves were simply known as “‘āina” or “moku(puni)” (places that were cut off from other lands by the sea).[31] A single island, standing alone, was known as a “moku kele i ka wa‘a” (an island that needed to be sailed to).[32] Two islands in close proximity to one another were referred to as “mokulua,” while an archipelago, such as the Hawaiian Islands, was known as “pae ‘āina” or “pae moku.” Islands close to one another were known as “‘āpapa ‘āina” or “‘āpapa mokupuni,” while scattered islands were known as “‘āluka” or “makawalu.”[33]

Islands were further divided into smaller land sections. By demarcating land divisions, ali‘i delineated the lands under their rule. Ancestral accounts credit Kālaihaohia—a kahuna living during the time of Kaka‘alaneo, a famous Maui ali‘i—as the first to carve Maui into districts, subdistricts, and even smaller divisions.[34] In keeping with this land system, the ali‘i nui of the island appointed an overseer to rule over each of the districts, subdistricts, and smaller subdivisions. By imposing human-made boundaries on the landscape, Kālaihaohia engaged in the production of place, inscribing it with meaning and effectively mapping the landscape.

In practice, however, creating manmade boundaries proved to be complicated. Land division practices were not clear-cut, varying from place to place. Even the most reliable scholars of the nineteenth century contradicted one another, suggesting that variation in naming practices occurred from place to place. The terms "'āpana," "kalana," "mokuoloko," and "'okana" (sections that had been cut off) seem to be sources of great confusion.[35] Both Kamakau and Malo agreed that islands are known as "mokupuni" and that islands or lands are known as "moku" or "'āina." Islands were further divided into districts. According to Kamakau, the largest land division of an island was the moku'āina, sometimes known as a kalana;[36] Malo used mokuoloko and 'āpana interchangeably in reference to the largest land division of an island.[37] Kamakau's and Malo's understandings of the terms 'okana and kalana differ as well. Kamakau states, "O ka okana, he mau mahele ia iloko o ka mokuaina a me ke kalana" ('Okana is a smaller land division than a kalana or a moku'āina).[38] Kamakau's use of kalana in reference to some large districts of land is likewise supported by records appearing in the *Buke Mahele*.[39] Cartographer W. D. Alexander's "A Brief History of Land Titles in the Hawaiian Kingdom," also written in the nineteenth century, provides a Maui example aligned with Kamakau's assertion. Alexander states, "On Maui there are some sub-districts called *Okanas*, of which there are five in the Hana district, while Lahaina is termed a *Kalana*."[40]

Pukui and Elbert disagree with Kamakau, defining a kalana as a "division of land smaller than a *moku* or district; county."[41] S. Mokuleia, a Kanaka writing in the 'ōlelo Hawai'i newspaper *Ka Nupepa Kuokoa* in 1868, supports Malo's assertion by stating, "Ina i mahele ia ka moku o loko i mau apana hou, he okana ia a he kalana kekahi inoa, a he poe maloko oia okana, a ua mahele ia i mau ahupuaa hou" (If the mokuoloko were subdivided, they would be called 'okana or kalana, and within an 'okana, it would be further divided into ahupua'a).[42] To add further confusion, R. D. King, principal cadastral engineer for the Survey Department of the Territory of Hawa'i, contradicted both Kamakau and Malo by suggesting that some ahupua'a were called kalana, stating that Kahakuloa, Olowalu, and Ukumehame—three recognized Maui ahupua'a—were sometimes referred to as kalana.[43]

TABLE 1 Understanding of land terms according to different sources

Source	Island	Large District	Smaller District	Subdistrict
Kamakau	moku mokupuni	moku'āina (sometimes kalana)	'okana	ahupua'a
Malo	moku mokupuni	mokuoloko='āpana	'okana (sometimes kalana)	ahupua'a
Pukui & Elbert	moku moku'āina mokupuni	moku	kalana 'okana	ahupua'a

By most accounts, the ahupua'a was the next largest land division. "Ahupua'a" is defined by Pukui and Elbert as a "land division usually extending from the uplands to the sea, so called because the boundary was marked by a heap (*ahu*) of stones surmounted by an image of a pig (*pua'a*), or because a pig or other tribute was laid on the altar as tax to the chief. The landlord or owner of an *ahupua'a* might be a *konohiki*."[44] C. J. Lyons states, "The Ahupuaa however was by no means any measure of area, as it varied in size from one hundred to one hundred thousand acres."[45] A popular notion is that all ahupua'a stretched from the mountains to the sea as pie-shaped wedges; however, as Kanaka scholar Donovan Preza points out, "This description has made the term ahupua'a synonymous with valleys. While this definition may generally be true for the windward sides of the islands and the island of Kaua'i, there are many examples where this generalization is not true."[46]

Valley walls were not the only natural features on the landscape used to define boundaries. Interestingly, the boundaries of all but one moku of East Maui were demarcated by a single rock on top of Haleakalā.[47] The rock had at least two names: Pōhaku 'Oki 'Āina (rock that divides the land) and Palaha (flat). According to Kanaka photographer and scholar Kapulani Landgraf in *Nā Wahi Kapu o Maui*, "Eight of the nine traditional land divisions of east Maui radiate from Pālaha, a large rock on the northeast edge of Haleakalā Crater: Kīpahulu, Hāna, Ko'olau, Hāmākua Loa, Kula, Honua'ula, Kahikinui, and Kaupō."[48] In this way, the land divisions of this area may be compared to a he'e (octopus). If one imagines a he'e situated on the top of Haleakalā and extending over East Maui, the tentacles could serve as land dividers.

Often, many 'ili 'āina were found in a single ahupua'a, although small ahupua'a were not always subdivided.[49] The term "'ili 'āina" was often shortened to "'ili."[50] There were two types of 'ili 'āina: lele and kū. Lele were land

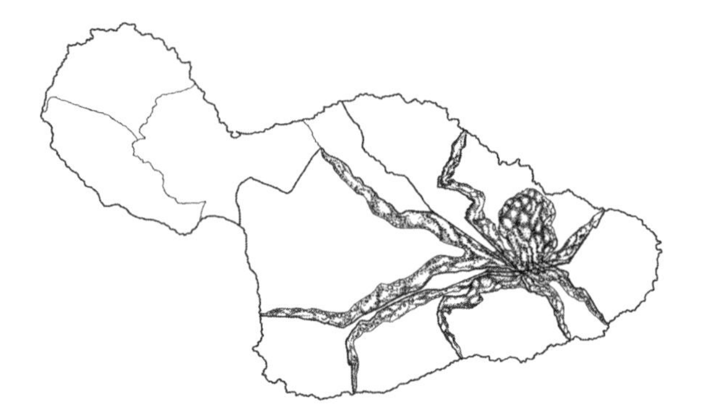

The tentacles of a heʻe demarcate the boundaries of the eight moku radiating from Pōhaku ʻOki ʻĀina

divisions tied to an ahupuaʻa politically and economically but not physically or geographically. The term "lele" means "to jump." Lele were separate lands that "jumped" over land and/or ocean depending on their location to be connected to another ahupuaʻa. These lele allowed for the equitable distribution of resources, thereby granting makaʻāinana access to resources that would not otherwise be found within the geographical confines of a particular ahupuaʻa. An aliʻi ʻai ahupuaʻa or konohiki oversaw the makaʻāinana of an ahupuaʻa, including those residing on lele. The konohiki, however, did not necessarily have authority over a second type of ʻili, one known as an ʻili kūpono (independent ʻili), sometimes abbreviated as kū. ʻIli ʻāina were in turn subdivided into moʻo ʻāina that were further divided into paukū ʻāina.[51] Within paukū ʻāina were even smaller land divisions known by various names, including one kōʻele,[52] kīhāpai, hakuone,[53] kuakua,[54] hakupaʻa, mālua, nanaʻe, kīpoho(poho), puluwai, and pāʻeli.[55]

Land-Sea Continuum

The ahupuaʻa system was a land-sea continuum; the ocean was an extension of the land, and the land was an extension of the sea.[56] Therefore, Kānaka were able to indicate location using the ocean and mountains as reference points. A person traveling upland was said to be going "ma uka," while someone near the ocean would be "ma kai." Additionally, Kānaka used prominent features on the landscape as reference points. On the island of Maui, a person looking for Makawao might be told to travel ma uka toward the summit of

Haleakalā, ma kai of Kula, and ma uka of Hāliʻimaile. This simple method of mapping the location of someone or something was an ingenious way of delineating place for island dwellers who can easily recognize where ma uka and ma kai are on the landscape.[57]

Similarly, "ko kula uka" and "ko kula kai" referred to the resources of the upland slopes and the region near the ocean, respectively. Kula lands, such as Honuaʻula, Maui, were characteristically red soil expanses with few trees. Pili

Ma kai and ma uka perspectives

grass, herbs for medicinal purposes, kalo, poi (mashed, cooked kalo thinned with water), and ʻuala (sweet potato) were shared and distributed by kula uka to kula kai residents. Likewise, those living on kula kai areas would exchange limu (seaweed), ʻopihi (limpet), and iʻa (fish) with kula uka family and friends. Through this system of exchanging ko kula uka and ko kula kai or ko kahakai, the land and sea constituted one continuous resource that provided for all (ON 196–197).[58]

As an extension of the ʻāina, the ocean was not a barrier between islands; rather, it was a pathway connecting islands to one another.[59] Due to skilled navigators' profound understanding of ocean swells, cloud formations, bird flight patterns, star constellations, the rising and setting of the sun and moon, and other natural phenomena, navigators knew where they were relative to where they had come from and where they were headed. The ocean and the heavens were merely extensions of the islands from which the voyaging vessel came. The akua of the heavens and ocean, ancestral deities, and kūpuna continued to follow them, guiding them in the right direction and assisting them during trying times on their journey.[60]

Upon traveling to new lands throughout Oceania, Kānaka were able to utilize forms of performance cartography to map the lands to which they traveled.[61] These performances were so effective, in fact, that once a new land was reached, navigators could reconstruct their voyages, instructing other navigators. By following such a map that enumerated elements of the natural environment encountered on the journey, even a navigator who had not previously visited the land being described could find his or her way to a place thousands of miles away. This was the case of Mau Piailug, a master navigator from the island of Satawal in Micronesia who navigated the first voyage of the Hōkūleʻa in 1976. In spite of the fact that he had never sailed from Hawaiʻi to Tahiti before, he was able to successfully lead the crew to Tahiti relying solely on ancestral navigational techniques.

Oceanscapes

The indivisible nature of the ʻāina and the kai is evidenced by some of the poetic names attached to places on land. Ka pae ʻāina, for example, was poetically known as "Nā Kai ʻEwalu" (the eight seas) (ON 243).[62] Similarly, as previously noted, the island of Maui was nicknamed "Nā Hono aʻo Piʻilani" (the Bays of Pi'ilani) in deference to Pi'ilani, the ali'i nui who ruled the island (ON 243).

Kānaka also ingeniously extended techniques commonly seen on land to the ocean, such as building sea walls below the surface of the water. The ocean fronting Pi'ilanihale, the largest known heiau in ka pae 'āina Hawai'i, is lined with underwater walls to protect the ma kai area fronting the heiau from intruders. Opposing forces attempting to invade by canoe would run aground on the submerged walls. This thorough understanding of the ocean enabled Kānaka to map the ocean floor and to select the best spots for wall construction.[63]

Kānaka also mapped the sea by naming each part of the ocean, from the moana (deep sea) to the po'ina kai (place where waves break) and the 'ae kai (place where the waves wash up on the beach). Mapping the ocean was a practical way of locating one's resources. The names given to these oceanscapes were often based on the physical characteristics or resources available in that place. Kai 'ō kilo he'e, for example, was the place that he'e could be caught, while kai hī aku was where aku (bonito) were plentiful.[64] Kānaka even named underwater features such as ko'a (fishing spots where fish were fed, raised, and harvested) and reefs.[65] Through the process of naming, places on the land and even those at sea were mapped and recorded in people's memory banks for future reference. Being able to mentally map a location was largely a product of enjoying an intimate relationship with a place. Women, for example, were generally the gatherers of limu.[66] They intimately understood the shoreline and, through experience, mapped the best places to gather different species of limu and shellfish.

Traveling until the 'āina could not be seen required another layer of understanding the oceanscape. Reference points emanating from the 'āina could no longer be used, and navigators had different maps than fishers, allowing them to successfully voyage over vast distances. These multilayered maps were necessary for carving out the oceanscapes for gathering purposes and for creating niches in which different people were masters of their trades. A navigator might be revered for his or her ability to traverse across the ocean with ease; however, the navigator would not necessarily excel at the various fishing techniques needed to sustain the crew. In this way, Kanaka maps were used to define physical locations of places as well as to create places in society for people.

From the heavenscape through the landscape and on to the oceanscape, place was a multidimensional continuum. The various layers of place were interrelated and directly impacted one another. Kānaka understood, for example, that pollution of streams affected marine life. They also recognized that deforestation of well-established rain forests changed the environment, resulting in less rainfall; thus, they remarked, "Hahai nō ka ua i ka ulu lā'au (rains follow the forest) (ON 50).[67]

Because the world was so vast, Kānaka related to place from their own perspectives within the continuum. As previously mentioned, the phrase "Nā Kai 'Ewalu" normally referred to the seas between the eight major islands, but for a person from West Maui, this same phrase referred to the eight channels seen from Lahaina.[68] Kanaka scholar Kaleikoa Ka'eo offers another interpretation, asserting that nā kai 'ewalu does not mean "eight seas"; rather it references the eight points of a compass to include all places within ka pae 'āina Hawai'i.[69]

Similarly, 'ākau and hema made reference to the north and south cardinal points, respectively; however, in conjunction with one's body, the same terms applied to the right and left sides, respectively. As long as a person faced west, 'ākau was both north and right, just as hema dually meant south and left. As soon as a person's orientation changed, however, so too did their sense of direction. No longer was 'ākau synonymous with *both* the north cardinal point and right side of a person's body. In this situation, one's sense of direction was based on one's own position within the continuum at any given place and time.[70]

Places were also related spiritually through their divine connections. Previously, cosmogonic genealogies linking ka pae 'āina to the akua were discussed. Not mentioned, however, were the kino lau found throughout the environment. The mapping of place as a continuum was so inclusive that it extended to plants and animals. While the highest atmospheres of the heavens were beyond reach, Kānaka honored the homes of the akua by studying the intricate layers of the land and recognizing earthly manifestations of the akua. In this way, Kānaka were able to relate to the akua through the birds that soared in the skies, the animals that roamed the 'āina, and the fish that swam in the depths of the ocean. For example, Lono, the god of fertility, was seen in the skies in the form of dark clouds. From the 'āina, he would rise back up to the heavens as steam billowing from active volcanoes. On land,

Lono would manifest himself as a pig, hinupua'a (a type of banana), or a kukui tree. In the ocean, he would take the form of the kūmū, 'ama'ama fish, or limu līpu'upu'u.[71] Through kino lau, the akua were able to manifest themselves in different realms, reinforcing the notion that the heavens, land, and ocean were intertwined and related.

Mapping Maui

Because Kanaka mapping techniques were often relative to one's location at any given place and time, further discussion of these practices may be enhanced by narrowing the discussion to a single island in ka pae 'āina Hawai'i. Maui, the second largest of the major twelve Hawaiian Islands, is a double island: an isthmus joins the Kahālāwai and Haleakalā mountains. The smaller neighboring islands of Moloka'i, Lāna'i, and Kaho'olawe were politically and geographically linked to Maui via ancestral land divisions. Lāna'i was a kalana of Maui and Kaho'olawe was an ahupua'a of Honua'ula.[72] Collectively, Maui, Moloka'i, and Kaho'olawe were known as Maui Nui. The United States Geological Survey describes Maui Nui as "a prehistoric Hawaiian Island built from seven shield volcanoes. Nui means 'great/large' in the Hawaiian language. 1.2 million years ago, Maui Nui was 14,600 square kilometres (5,600 sq mi), 50% larger than the present-day island of Hawaii. As the volcanoes subsided and eroded, the saddles between them slowly flooded, forming four separate islands: Maui, Molokai, Lanai and Kahoolawe by about 200,000 years ago."[73] The ali'i who ruled Maui often ruled over these islands as well, although the ali'i of O'ahu and Maui often struggled for control of Moloka'i.[74] At the time of Captain Cook's landing, for instance, Kahahana ruled the islands of O'ahu and Moloka'i, while Kahekili ruled the islands of Maui, Lāna'i, and Kaho'olawe.[75] However, the grouping of all these islands in the epithet "Maui Nui a Kama" demonstrates that ancient Kānaka recognized the geological connection of these islands to one another.

Ancestrally, Maui had twelve moku. It also had five major centers of population: Kahakuloa, Nā Wai 'Ehā (the four waters), the southwest coast of West Maui extending from Olowalu to Honokōhau,[76] Ke'anae and Wailuanui, and Hāna.[77] Interestingly, at times Nā Wai 'Ehā (comprised of Waikapū, Wailuku, Wai'ehu, and Waihe'e) were four 'ili kūpono.[78] In fact, during the Māhele,

Waihe'e and Wai'ehu were considered to be located in Pū'ali Komohana (west of the Maui isthmus), while the adjacent ahupua'a of Wailuku and Waikapū were documented as being in Nā Poko (the smaller land divisions).[79] All four ahupua'a were ancestrally referred to collectively as Nā Wai 'Ehā because all four begin with the word "wai." This fertile region was known for its abundance of water and wetland lo'i.

In 1859, the land divisions of Maui were revamped, reducing the number of moku to four for taxation, educational, and judicial purposes; no longer were Moloka'i and Lāna'i part of the Maui moku.[80] The Civil Code of 1859, section 498, made the following revisions: "The islands of Maui, Molokai, Lanai and Kahoolawe, shall be divided into six districts as follows: 1. From Kahakuloa to Ukumehame, including Kahoolawe, to be called the Lahaina district; 2. From Waihee to Honuaula inclusive, to be called the Wailuku district; 3. Kahikinui, Kaupo, Kipahulu, Hana and Koolau, to be called the Hana district; 4. Hamakualoa, Hamakuapoko, Haliimaile, Makawao and Kula, to be called the Makawao district; 5. Molokai; 6. Lanai."[81]

King describes the changes more succinctly by discussing the changes in land divisions according to the names of the moku:

> In the Maui group, the twelve ancient districts of the island of Maui were reduced to four by combining Kaanapali with Lahaina, adding to it at the same time the island of Kahoolawe; by retaining in the Wailuku district the *ahupuaas* of Waihee, Waiehu, Wailuku and Waikapu but for some reason adding to it the ancient district of Honuaula though separated from Wailuku by the intervening district of Kula; by consolidating the districts of East Maui into one and the districts of central East Maui into another. The names of Kona and Koolau on Molokai were dropped and that island created into a separate single district called Molokai district, with the island of Lanai forming another separate district.[82]

Chapter XXIII of the Laws of 1890 once again revamped the land divisions of Maui, placing Kaho'olawe in the moku of Wailuku rather than Lahaina. In 1890 and 1894, for police court jurisdiction reasons, some revisions were made in Hāna.[83] It was not until fifteen years later, however, that large-scale changes occurred via Act 84 of the Session Laws of 1909.[84]

The Fluidity of Place

Kānaka mapped places by constructing and reconstructing their boundaries. Both tangible and intangible boundaries defined place. Prominent natural features and man-made altars and rock walls often served as clearly visible landmarks. Kūkulu ʻehoʻeho (mounds of stones) were employed so that occupants knew what resources were within their jurisdiction and to which aliʻi they were indebted.[85] Banks between loʻi were functional, allowing people to traverse easily between gardens while simultaneously dividing one farmer's loʻi from that of a neighbor. Moʻo (narrow strips of farmed land) and iwi (mountain ridges) often served as the dividing line between ahupuaʻa and ʻili ʻāina.[86] Even imaginary boundaries, such as those demarcated by the rising and setting of the sun, moon, and stars, as well as the intersecting of two or more landmarks at sea, defined place.[87]

Although Kānaka knew the parameters of their homesteads and gardens from working and living in the same place for generations, the values of generosity, love, and compassion permitted passage through known boundaries. As Kanahele suggests, "outside the sacred precincts, ordinary Kānaka allowed each other a good deal of freedom to enter, cross, and leave their homesteads. But this courtesy, we assume, was based on the principles of reciprocity and hospitality which, in effect, softened the somewhat hard attitudes implied in the concept of territoriality."[88]

Boundaries of profane places were generally fluid and constantly being redefined and kālai ʻia (carved out). In fact, regions within the heavenscapes, landscapes, and oceanscapes did not have clear-cut, distinct boundaries. Kānaka did not, for example, rely on specific altitudes to delineate the margins of the kua mauna and the kua lono. Land regions were often defined by the type and size of vegetation growing within particular zones and the characteristics of each expanse. The boundary dissecting two regions was not a single concrete line; rather, buffer zones were more frequently employed.[89] Even during the period of the Māhele, in which identifiable "boundaries" were necessary, they remained somewhat abstract at times, as the following example illustrates: "The ancient Hawaiian method of determining the dividing lines between hillside and valley property is illustrated in testimony before the Land Commission in 1848 to affix the boundary between Kewalo and Kaimuohena. 'The dividing line between them is where a stone would stop when rolled down the ridge. Kewalo is any place where a stone running down would stop and below where a stone would stop is Kaimuohena.'"[90] In this example, no survey reference points are provided. Moreover, one could argue that each time a rock was rolled, it would stop in a different location, making for a very fluid boundary.

Boundaries for places are sometimes permeable. A canoe, for example, is usually associated with the ocean; nevertheless, for the canoe, the land and ocean are inseparable entities. The canoe transcends such boundaries by virtue of its interconnectedness to both, traveling over the ocean and yet usually being pulled ashore for storage. Similarly, the canoe links places by ocean. The Kūkahiko family of Mākena, for example, continues to travel back and forth between Maui and Kaho'olawe on fishing expeditions, as our kūpuna have done for generations. In this way, the family has been able to maintain a strong relationship to our kulāiwi as well as preserve the bond between the island of Kaho'olawe and Mākena on the island of Maui, two land sections in the moku of Honua'ula, Maui.[91]

Ancestral Kānaka understood the fluidity of profane place and did not try to compartmentalize it into separate distinct categories; rather, boundaries flowed one into another. Their thinking was seemingly more inclusive. Phrases such as "mai kekahi kapa a kekahi kapa aku" (from one place to another and everything in between) were a very inclusive way of perceiving the world around them. The same could be said of various regions. They were a continuum, at times even overlapping, creating areas that were not black or white but gray.

As Kānaka became influenced by the Western world and the concept of private landownership, boundaries took on new meaning. As a result of the Māhele, surveyors delineated parcels with precise boundaries. Even today, land title searches rely on the assigned tax map key number and boundaries as recorded by the Bureau of Conveyances. How did Kānaka cope with this new system of boundary making?

One such way was through the creation of hui (joint ownership) lands that entitled people to an undivided share in a parcel. For some Kānaka, this made the transition between communal land stewardship to private landownership more comprehensible, allowing people to move toward private landownership while simultaneously maintaining the more permeable boundaries of ancestral landownership. As Andrade notes,

> The *hui* was one of many associations formed by *maka'āinana* throughout the islands to buy land in post-Mahele/Kuleana Act times. Although not well known, the story of this movement gives voice to a short-lived counteroffensive against the ongoing alienation of land and dispossession brought on by the new regime of private property and real estate. Although the Kuleana Act entitled subsistence gathering, access, and water use rights to *maka'āinana* awardees, in many cases, this was not enough to guarantee the means of survival provided by their former, cooperative practices: customary rights to the resources of the *ahupua'a* and the traditional system of reciprocal responsibility.[92]

Ironically, today, descendants of those who originally joined hui lands, accustomed to communal land tenure, are now finding the same system that their ancestors relied on for continuity to be plagued with problems. Since the Māhele in 1848, people residing in Hawai'i have become used to clearly delineated boundaries for private landownership. Hui lands, though, do not have clearly defined borders because co-owners share an undivided interest in the hui. Without boundaries that can be surveyed and measured to demarcate individual interests within the hui, owners often have difficulty securing mortgages because loan officers cannot use the property itself as collateral, should the client fail to make payments.[93] Moreover, with the rise of drug use in some areas, another concern arises: should someone within the hui conduct illegal activities on the premises, the entire hui could be confiscated, not just that person's interest, because everyone's interest is undivided. Unlike parcels with tax map key numbers that have very rigid boundaries, hui lands have more blurry boundaries. In fact, the only boundaries that are clearly defined are those that outline the entire hui land.

Wahi Kapu: Distinct Boundaries

Kapu places had much more mana than their profane counterpart, and they were therefore inscribed with distinct and rigid boundaries. The physical presence of ali'i often defined kapu places. The mere fact that the ali'i, and the akua they worshipped, revered certain places bestowed mana on those places. Ancestrally, places such as heiau and royal residences, considered to be extremely kapu, were demarked by pūlo'ulo'u (insignia of chiefly kapu) and were guarded by royal attendants. In a society where people were sentenced to death for coming into contact with so much as the shadow of an ali'i nui, clearly defined boundaries between the sacred and profane were necessary.[94] Delineating boundaries on the landscape allowed for maintaining control and order over places. By increasing their place, ali'i enhanced their mana. Ali'i therefore warred with one another in order to both increase their own holdings and to defend what they had already acquired.

Conclusion

Ancestral Kānaka recognized the connection between the heavens, lands, and oceans and how all three were interconnected and interdependent upon one another. In spite of the interwoven nature of the sky, land, and sea, however, Kānaka of ancestral times did not have a term that directly translates to what we have come to know today as "environment." Rather, the *Hawaiian Dictionary* offers two phrases that approximate the notion of environment: (1) "'ano o ka nohona" and (2) "nā mea e ho'opuni ana." 'Ano o ka nohona refers to the nature of one's relationship to one's surroundings or places. Nā mea e ho'opuni ana relates to everything that surrounds or encircles a person.[95]

The fact that an 'ōlelo Hawai'i term for environment did not exist in ancestral times may suggest that Kānaka had a more intimate relationship with their surroundings than we do today because coining a single term for everything in their environment was insufficient for their needs. Instead, they engaged in the more intimate practice of naming the individual parts of the environment, including the strata of the heavens, stars in the sky, regions on the landscape, and depths of the ocean. By doing so, spaces were transformed into places; each part of nā mea e ho'opuni ana was recognized and mapped.[96]

Chapter 4 Kanaka Performance Cartographies

Ancestrally, Kānaka did not have maps in the form of written representations of the world; instead they utilized "performance cartographies" to reference their constructed places, legitimize their existence, and reinforce their legacies.[1] Such cartographic representations were expressed in many ways in ka pae moku, including but not limited to the following: inoa ʻāina (place names), mele (songs), hula (dance), ʻōlelo noʻeau (proverbs), māhele ʻāina (land divisions), moʻolelo (historical accounts), moʻokūʻauhau (genealogies), kaulana mahina (moon calendars), hei (string figures), and hoʻokele (navigation).

The modes of expression utilized in ancestral Kanaka performance cartographies functioned like maps, referencing spatial understandings and features, yet they were largely oral in nature.[2] Like other indigenous peoples, ancestral Kānaka were "able to relate their own sense of being, expressed in cosmology, genealogy, history, and lived experience, in the oral map that is laid on the landscape."[3]

Via performance cartographies, Kānaka connected places of significance to one another—no matter how distant they might be on the landscape. *Nā Kuahiwi ʻElima,* for example, is a mele written about five peaks. Four of the mountains are found on the island of Hawaiʻi, while the fifth is Haleakalā on the neighboring island of Maui. Another composition, *Nā Wai ʻEhā,* links four fertile lands on Maui into one seamless region. This chapter explores ancestral Kanaka "cartographic" practices and how these practices contributed to the construction, maintenance, and contestation of Kanaka place in ancestral times.[4]

Waihona Noʻonoʻo: Place Holds Memory

Place holds memory for people who share an intimate relationship with their environment.[5] Those who are born and raised on their kulāiwi tend to cultivate a strong connection to their one hānau. Even people who leave their birthplace often retain vivid memories of their kulāiwi later in life; thus, places serve as mnemonic devices, helping people to record their past. For kūpuna who have not returned to their homeland in decades, just the sight of a place in a picture or postcard can jog the memory.

All memories are embodied and grounded in place.[6] As Tewa scholar Gregory Cajete asserts, "It is the landscape that contains the memories, the bones of the ancestors, the earth, air, fire, water, and spirit from which a Native culture has come and to which it continually returns. It is the land that ultimately defines a Native people."[7] Memories of one's childhood stomping grounds, school, and swimming spot are all anchored to specific places; therefore, memories are either of a place or of an event that occurred in a place. Embedded in such memories is also one's relationship with those places and events; conversely, the bond that one establishes with a place and event leads to the construction of memories.

To truly know a place is to be able to recite its stories. As a result, place-based memories may reveal themselves in place-specific moʻolelo. For Kānaka, these moʻolelo are often family treasures handed down to succeeding generations. Some families are still able to perform the same walking oral histories of their kulāiwi as their kūpuna once did. As they walk along the coastlines, valleys, and other places on the landscape, they vividly recall the moʻolelo and place names of the area and nostalgically reminisce about how the place once was.

At a family reunion in Mākena on the island of Maui, I met a ninety-four-year-old relative who paid for all of his children, grandchildren, great-grandchildren, and great-great-grandchildren to fly to Maui to attend what would be his last family reunion. With all of his descendants surrounding him, he fondly recollected the memories he had as a child growing up on his kulāiwi. With tears of joy at being reunited with his kulāiwi and tears of sadness knowing that it would likely be his final visit home, he affectionately recited the place names, pointed out the former house lots of relatives, and shared favorite past times associated with the area.

Mele Wahi Pana: Chanting the Landscape

To know a place is also to be able to chant the landscape through poetry. As eloquent orators, master chanters, and prolific composers, ancestral Kānaka utilized mele to "map" their relationships to places.[8] Through mele and kaona, kupa o ka ʻāina described the beauty of the mountains, the sound of the ocean, and the glow of the sunset. Being able to hoʻopōhaku (to live in the same place for generations) sparked a sense of pride and served as inspiration for new compositions.[9] Inspired by their intimate knowledge of the landscape, haku mele (musical composers) masterfully enumerated the ʻōlelo noʻeau, winds, rains, and place names associated with the places glorified in their compositions.

Mele wahi pana (songs honoring storied places) were common in ancestral times in ka pae ʻāina Hawaiʻi.[10] Today, contemporary haku mele continue this tradition by composing mele that venerate our cherished places. Made famous in song is the phrase "Ua noho au a kupa" (I have resided until well acquainted with this place).[11] Similarly, Kānaka familiar with a place fondly exclaim, "Ua hele au i kēia mau kuahiwi a lewa" (I have traveled these mountains so extensively that I know every nook and cranny). Composing a mele in honor of one's homeland is perhaps one of the greatest tributes a person could bestow upon their kulāiwi.

Hanohano ʻo Maui

Mele serve as a form of geography and performance cartography for Kanaka places.[12] Deeply embedded in mele are references honoring place. *Hanohano ʻo Maui,* composed by Kumu Hula (master of the art of hula) and Kumu ʻŌlelo Hawaiʻi (Hawaiian language teacher) Kahikina de Silva, for example, celebrates the names and characteristics of various places on Maui. Such a composition honors each of the places noted in the mele and etches the characteristics described in the memory of the singer and audience. Even someone who has not previously been to Maui will learn of the chilly waters of ʻĪao and the warmth of Lahaina. Mele honoring places have the ability to transport people from their current location to distant places simply by virtue of eloquent poetry and creative imagery. Such songs are an important way of mapping an island. Through the use of place names as mnemonic devices, even someone unfamiliar with an island is able to travel around the island and know where she is.[13]

Hanohano ʻo Maui

Hanohano ʻo Maui i ka lei loke	Maui is honored with the rose lei
A he nani hiwahiwa kū hoʻokahi	An esteemed beauty that stands alone
ʻAkahi hoʻi au a ʻike maka	I have just seen for myself
I ka wai huʻihuʻi aʻo ʻĪao	The icy waters of ʻĪao
ʻO ka nehe a ke kai i Puamana	The rumbling of the sea at Puamana
Kahi o nā hana leʻa aʻo Kananaka	Where Kananaka played joyously
Kuahiwi kilakila ʻo Haleakalā	Haleakalā is a majestic mountain
me ka ua huʻihuʻi koni i ka ʻili	With the cold rain that tingles on the skin
E aho nō ʻoe e komo mai	It's better that you enter
I ka ʻāina pumehana o Kāʻanapali	The warm land of Kāʻanapali
Mehana hoʻi i ka malu aʻo Lele	Warmed in the protection of Lahaina
Pio ʻole ke kukui i ka ua ʻUla	Where light is unextinguished by the ʻUla rains
Puana ka inoa a e lohe ʻia	Tell the name that it may be heard
ʻO Maui aʻo Kama nō ē ka ʻoi	Maui of Kama is indeed the best

(SOURCE: Composed by Kahikina de Silva, from liner notes for Kealiʻi Reichel's compact disk, *Kamahiwa*)

ʻAuhea Wale Ana ʻOe

Another example, *ʻAuhea Wale Ana ʻOe,* composed by Nakulula, is a mele wahi pana, mele hoʻohanohano (honorific chant), and mele aloha (love chant). By virtue of eloquent poetry, the haku mele intertwines place names and feelings of love and admiration into a single mele. In the final line, Nakulula acknowledges that the mele was composed for Mākaʻitūana.

In this composition, the haku mele uses performance cartography to map the winds, rains, and other characteristics of the places being referenced. Through the names of the winds and rains, those listening to the mele are able to envision and feel the natural elements of the places alluded to in the mele. The name "ʻUlalena" (yellowish red), for example, implies that this is a yellowish-red hued rain and wind. The composer's reference to billowing clouds places ʻAwalau in the uplands. In Makawao where the ʻŪkiukiu rain falls, the wind blows gently, while the plains of Kamaʻomaʻo experience sudden bursts of showers, as is consistent with the characteristics of the Nāulu rain.

Although each carefully selected place, wind, and rain name reveals a great deal about the natural environment, through the art of kaona and layering thoughts, haku mele are able to poetically allude to other aspects of life as well. The different winds and rains, for example, may be compared to various acts and emotions associated with human intimacy. The name "ʻŪkiukiu," for instance, is derived from the word "kiu" (to spy). It is very probable that the haku mele of this song had every intention of using the name of this rain to suggest that someone was spying, perhaps over a prospective lover. While meaning making is in the mind of the listener, only the haku mele knows for certain her intended meaning(s). After all, Kanaka poetry promotes multiple meanings rather than implying that there may be a singular true meaning.

ʻAuhea Wale Ana ʻOe

ʻAuhea wale ana ʻoe ē E ta ua ʻUlalena	Where are you ʻUlalena rain
Kāhiko maila i uka ē I ka nani o Piʻiholo	Adorning the uplands, Amid the beauty of Piʻiholo
Ua like me ka ʻōpua ē Noho maila i ʻAwalau	Like the clouds Residing at ʻAwalau
ʻAu aʻe nei ka manaʻo ē E pili me ke aloha	My thoughts flow For my beloved
Aloha ʻo Makawao ē I ka ua ʻŪkiukiu	Beloved is Makawao In the ʻŪkiukiu rain
He kiu na ka Nāulu ē I ke kula o Kamaʻomaʻo	An observer for the Nāulu showers At the plains of Kamaʻomaʻo

ʻO ka loa kaʻu i ana ē	I measured the expanse
I ka ʻoni o ka lihilihi	That stirred briefly
Ilihia iho nei loko ē	A thrill taken over
I ka ukana o ke aloha	From the love carried
Haʻina mai ka puana ē	Let the story be told
Mākaʻitūana he inoa	A name song for Mākaʻitūana

(SOURCE: Composed by Nakulula, from liner notes for Kealiʻi Reichel's compact disk, *Kamahiwa*)

Koali

The following mele describes many aspects about Koali, a place in the moku of Hāna on the mokupuni of Maui.[14] First, we learn of two places within Koali, named Kānewai and Wailua. Because of references to the ocean breaks and sea sprays, it is evident that both places are along the shore. Second, we are introduced to some of the plants that can be found in Koali. The rose, ʻiwaʻiwa fern, niu (coconut), and fragrant līpoa (a type of seaweed) are enumerated as flora of this place. Third, in the way of fauna, the ʻiwa bird is noted for soaring along the shoreline and cliffs. Fourth, natural features such as characteristics of the shore and cliffs are revealed. Fifth, a sense of national pride for the Kingdom of Hawaiʻi is alluded to by the hae Hawaiʻi (Hawaiian flag) reference.

Through the careful word choices selected by the eloquent haku mele, a great deal of information about Koali is exposed in this mele aloha. It is undoubtedly a song of fondness for the place, and perhaps it is also a song of affection for a lover that is only alluded to in kaona. *Koali* is an excellent example of an honorific mele that highlights and maps important aspects of this place.

Koali

He aloha no Koali	Koali is beloved
I ka nalu haʻi mai a ʻo Kānewai	for its breaking waves at Kānewai
Me ka līpoa paoa ʻala (pā ʻaʻala)	with the fragrant scent of the līpoa
Anu makehewa kēlā	it is extremely cold

Ua hoʻi i ka ulu ia hale rose	Returning to the house of the rose
I ka hale lau o ka ʻiwaʻiwa	to the house of the many ʻiwaʻiwa ferns
Me nā ʻiwa e walea ai	with the ʻiwa birds that relax
Kīkaha mālie i ka laʻi	gliding gently in the calmness
ʻO ka mahina hiki aloalo	The moon rises to its zenith
Pōhina wehiwehi i nā pali	misty in the cliffs
Me ka līpoa ʻala onaona	with the fragrant scent of the līpoa
Koni māʻeʻele i ke kino	that pulsates the body until numb
Nani Wailua i ka ʻehu o ke kai	Wailua is beautiful in the sea spray
I ka holu nape o ka lau o ka niu	in the swaying of the coconut fronds
Kōwelo haʻaheo Hae Hawaiʻi	your Hawaiian flag waves proudly
Kū kilakila i ka laʻi	standing majestically in the calm
Haʻina ʻia mai ana ka puana	The story has been told
I ka nalu haʻi mai a ʻo Kānewai	about the breaking waves of Kānewai
Me ka līpoa paoa ʻala (pā ʻaʻala)	with the fragrant scent of the līpoa
Anu makehewa kēlā	it is extremely cold

(SOURCE OF LYRICS: Kimo ʻAlama Keaulana's mele collection)

KANANAKA

The mele that follows was written for a moʻo (demigod found near water sources, often compared to mermaids in a contemporary context) named Kananaka. In this mele, many aspects of the natural environment are recorded. In the first verse, the Maʻaʻa wind that carries the scent of the līpoa is mentioned. Because this wind is famous in Lahaina, without even mentioning the place name the mele is "placed" on the landscape. Someone seeking līpoa for a family meal or healing practices would learn that Lahaina is an ideal place to gather this resource.

The sand dunes and surfing spot of Kananaka once again map the natural features of Lahaina. The island of Maui is famous for its sand dunes, and Lahaina is no exception. Perhaps this mele draws attention to these features in order to suggest that Lahaina was a stomping ground of the aliʻi, which is not at all surprising because Lahaina was the former official capital of ka pae ʻāina Hawaiʻi. Moreover, Kānaka often buried their dead in sand dunes in ancestral times; perhaps these sand dunes were the final resting places of aliʻi.[15]

Furthermore, references to the ali'i sport of surfing once again establish the home of Kananaka as a place that was once frequented by ali'i.

Kananaka

'O ka pā mai a ka Ma'a'a	The blowing of the Ma'a'a wind
Halihali mai ana lā i ke 'ala	carries the scent
Ke 'ala onaona o ka līpoa	of the fragrant līpoa
Hana 'oe a kani pono	you take it until satisfied
hui	chorus
Nani wale ia pu'e one	The sand dunes are beautiful
I ka nalu he'e mai a'o Kananaka	in the surfing spot of Kananaka
Kahi a mākou i he'e ai	the place that we surf
I ka 'ehu'ehu o ke kai	amid the sea spray
'O ka mahina hiki aloalo	The moon rises to its zenith
Ho'ola'ila'i ana lā i nā pali	relaxing amongst the cliffs
Pōhina wehiwehi i ke onaona	it is misty and lush with the fragrance
Koni ma'ele'ele i ke kino	that pulsates the body until numb[16]

(SOURCE OF LYRICS: Wilcox et al., 2004)

Hula: Bringing the Words to Life

The art of hula is another form of performance cartography. Through hula, dancers are able to bring the words of a mele to life. By watching a captivating hula dancer, the audience is able to visualize a song, and the elements of nature come to life via dance. The act of dancing a mele inoa in honor of an ali'i, for instance, demonstrates a person's respect and admiration for that ali'i. Mele recalling the superior leadership skills of an ali'i often end with a final kāhea (calling out the name) by hula dancers, once again acknowledging the ali'i for whom the song was composed.

Even in modern times, people celebrate dances of various famous ali'i, and through these forms of performance cartography Kānaka living many generations after the death of an ali'i are able to use these mnemonic devices to understand the past. Hula choreographed to mele inoa honor the life of beloved ali'i. Conversely, the names of ali'i not carried on through such performative cartographies are forgotten in due time, and their place in history is replaced by the memory of a more famous ali'i.

One such mele inoa that is widely known today is *'Au'a 'Ia E Kama Ē Kona Moku*. Composed for 'Aikanaka, the maternal grandfather of Lili'uokalani, by Keāulumoku, the same prolific poet who also composed the *Kumulipo, 'Au'a 'Ia E Kama Ē Kona Moku* is often used today to rally Kānaka together in opposition to forces that seek to diminish Kanaka rights.

'Au'a 'Ia E Kama Ē Kona Moku

Au'a 'ia e kama ē kona moku
'O kona moku ē kama e 'au'a 'ia
'O ke kama, kama, kama, kama i ka hulinu'u

'O ke kama, kama, kama, kama i ka huliau
Hulihia pāpio a i lalo i ke alo
Hulihia i ka imu o Kū kamaki'ilohelohe
'O ka hana 'ana ia hiki'i hulahula
Ka'a 'ia 'alihi a'o pōhaku kū
Me ka 'upena aku a'o Ihuaniani
Me ke unu o Niu-olani-o-La'a [sic]
'O Keawe 'ai kū, 'ai a La'ahia
Nana [sic] i hala pepe ka honua o ka moku
I ha'alē'ia i ke kiu welo ka pu'u kōwelo lohi a Kanaloa

Kama retains his lands
The lands that are retained by Kama
Kama, Kama, Kama, Kama of the highest rank

Kama, Kama, Kama, Kama of the changing time
Overthrown will be his foes, left lying face downward
Overthrown will be the imu of Kū with the sacred maki'ilohelohe cord
The cords that bind hulahula
Braided are cords of the anchor stone
And the bonito nets of Ihuaniani
And the temple of Niu-olani-o-La'a
And from Keawe, the sacred one
The one who ruled and made the island his own
His power rose to the summit of the hills, this is the powerful descendant of Kanaloa[17]

(SOURCE: Nona Beamer, *Nā Mele Hula*, vol. 2: *Hawaiian Hula Rituals and Chants*, 67)

Mele Ko'ihonua and Mo'okū'auhau: "Mapping" the Genealogical Connections

Through the *Kumulipo* and other mele ko'ihonua, the creation of ka pae moku unfolds. From space as a void to the unions of the gods, place is mapped in genealogical chants and other genera of mo'olelo passed down generation after generation. As you may recall from *Mele a Pāku'i,* for example, the birthing of the islands is enumerated as a genealogical chant. Wākea and Papa procreate, and Kahitikū, Kahitimoe, Ke'āpapanu'u, and Ke'āpapalani are born, followed by the islands of Hawai'i and Maui(loa).[18] Papa and Wākea each have children with other partners before reuniting. Together, they are parents of the islands: Kaua'i, Ni'ihau, Lehua, and Ka'ula.[19] Such mo'olelo map places genealogically, acknowledging the birth order of islands and geographically revealing their physical nature.

In the mele *'O Maui 'o Papa 'o Papaikanī'au,* the composer enumerates the genealogies of various Maui ali'i, the akua from which they descend, the winds and rains of their homelands, and the places they call home. This mele typifies poetry honoring ali'i. In order for an ali'i to have a place to stand and a right to rule, the connection to one's birthplace, ancestral lands, akua, and high-ranking kūpuna must be clearly established. Together, these building blocks form a strong foundation for ali'i.

'O Maui 'o Papa 'o Papaikanī'au

'O Maui 'o Papa 'o Papaikanī'au	It is Maui of Papa, Papaikanī'au
'O Hāhō [sic] o ka lulu kaupakapaka	Hāhō, [sic] of the many protected descendants
Nāna ia wahi 'o 'Īao	'Īao belongs to him
Na kā'e'a'e'a ia o laila	The ruler there
'O Kalākaua'ehuakama	Kalākaua'ehuakama
No ka pela ali'i o Wākea	From that chiefly line of Wākea
No loko o Haunaka ke li'i	From within the chiefly line of Haunaka
'O Kaualū o Lo'e	Kaualū of Lo'e
'O ke kūmaka 'alaneo	Witnessed clearly
'O Lā'ielohelohe o kala makua	Lā'ielohelohe from the old line of chiefs
'O naha kapu a Pi'ilaniwahine	The sacred "naha" status of Pi'ilaniwahine
Puka mai ka maka o ka hai Pi'ilani	Pi'ilani's progeny emerge
'O 'Olepau, 'o Kāulahea	'Olepau, Kāulahea

ʻO Kama ʻo Kamalalawalu [sic]	Kama, who is also known as Kamalālāwalu
No laila mai Keohokālole	Keohokālole is from this lineage
Nāna i hānau Kamakaʻeha	She is the one who gave birth to Kamakaʻeha
Ke liʻi nāna i kahiko o [sic] Maui lā	The aliʻi of the senior line of Maui
Kāhiko i Kekaʻa ka ua Nahua	The Nahua rain adorns Kekaʻa
He ua Nahua ua Lililehua [sic]	The rains of Nahua and Līlīlehua
Ua makaʻu pili ua Kauaʻula	Following is the ravaging rain, Kauaʻula
Ua noho i uka o wai a wai ʻehā	Residing upland in the region of the four waters
He ao ʻole ia nei he naʻaupō	Without enlightenment there is ignorance
He kiʻi ka hai nei a i waiho ai e ke ʻaʻe	Fetching the things that were left behind by the defiant one
Kiʻina e ke kuho ala naʻina e Kamanaʻo	Seeking with the intention to conquer
Kuhikuhi i waena konu o ke kanaka	Instructing the core of the people
Molale ia wahi akāka ke mana ia lā	The space becomes clear when invoked
Nānā ʻia e Līhau i luna	Seen by Līhau of the upper echelon
Waiho wale na Lono i ka mālie, e ala ē!	Left by Lono in the calm, arise!

(SOURCE: K. Kanahele, *Mauiloa*, 8)

Through the oral recitation of mele koʻihonua and moʻokūʻauhau, the names of ancestors were mapped. Kanikau (dirges), on the other hand, honored the lives of the people for whom they were composed. For aliʻi, such maps often recalled their acts of bravery and skill as strategists, leaders, and warriors. Landholdings were often enumerated to praise the accomplishments of the aliʻi. The political agenda was to demonstrate that the aliʻi was undoubtedly an akua who resided on earth and ruled over many lands.[20]

A kanikau composed in honor of Peʻapeʻa, an aliʻi of Maui, for example, begins with terms of endearment for the beloved chief. It continues by acknowledging his genealogy and hōʻailona, such as clouds in the sky, which are marks of a chief. The kanikau also maps the places that he frequented throughout his life. It closes by stating that Peʻapeʻa has been called home to the heavens. The composer mourns the loss of the beloved chief.[21]

He Kanikau no Pe'ape'a

He kanikau aloha keia,	This is a mourning chant,
Nou hoi la e Peapaemakawalu [sic],	For you, Peapaemakawalu [sic]
Kamakauahoa,	Kamakauahoa,
Kau mea 'loha i nalo aku la.	My beloved one who passed away,
O ku palena nui a Haho,	The large land division ruled by Haho.
O kama luaia o Palena,	The second child of Palena,
O Holaniku a Kaihe ka makua.	Hōlanikū of Kaihe was the parent
Ka ukali hope, ka puaa kau i ke aolewa,	The attendant, the pig in the clouds.
O Kaohelelani a Lono, na hoa hele,	Kaohelelani and Lono are the traveling companions,
Ka ukali o ka hope,	The attendant follows
O ka hookualana ana o kahi e	With failing strength from somewhere else.
Ekolu lakou e ahu nei,	Three of them gathered,
Haalele i ka moku i ka aina,	The ship left the land.
O Hana keia, akahi o loaa,	This is Hāna where we have just arrived,
O Hana, aina ua lani haahaa,	Hāna, the land of low heavens,
Lanakila nei o Kauiki, mauna i ka lani,	Ka'uiki is victorious, mountain in the heavens.
O Kapueokahi, o Mokuhono i kai o Kaihalulu,	Kapueokahi and Mokuhono of the ocean of Kaihalulu,
O Manianiaula, o Hamaalewa o Kauiki,	Māniania'ula, Hamaalewa and Ka'uiki,
Mauna i ka lani, ka mauna i ka paipai,	Mountain in the heavens, the mountain at great heights,
Hale o ka lani i hele aku la,	Home of the chief who passed away,
E o ia nei o Kamakauahoa o Kepanila.	Calling upon Kamakauahoa of Kepanilā,
O kau-hai-paku ka hoi	And also Kauhaipaku.
Hoi makani o Kamakauahoa,	Kamakauahoa returned in spirit,
Kau mea 'loha i nalo aku la,	My beloved one who passed away,
E uwe oe, e helu au o Kumukoa aku, o ka lani.	Weep you, I recite the merits of Kumukoa, of the chief.
I aha oe i welawela ai i punini ai oe.	What made you angry that you should go adrift,
Hele ka hoa o ke kaikunane,	The companion of my brother went.
O olua ia a Neau, mai loko o ka hale pupuu hookahi nei.	Both of you are of Neau, from the same womb,
Akahi no ka ke aloha,	A singular bond of love,
Ka paumako ia oukou,	Intense grief for you

E ahu i ka wai o Punahoa,
Nau ka e moe ke na wai,
Ka wai halana kiowai a ka ua,
I hookio ia e ka ua apuakea,
E ka makani koholalele nei.
O oe anei kahi anoai?
Ka ua wawahi i luna o ka hala.
Ka hala mai Akiu a Honokalani

Ka ulunahele hala o Akiola,
Ka'u mea 'loha i nalo aku la.
He aloha la ko'u e noho aku nei,
He maeele no ka lima ia oe,
Aloha ka-lani, e aloha ka-lani,
Aloha ka-lani i hele aku nei,
Ua ahi ka-lani,
Ua momoku ka ili,
Ua mea e ka lani, ua kino akua.
Ua kino lau, kino lau pahaohao.
Ua haona ke kino o ka lani i ke akua.
Ka lani, akua hou o Koolau,
I hoi i ka lulu o Kapueokahi.

Ka poe hanehane i Kaiakahauli,
Ka kini noho kahakai o Nanualele.

O ka lani, ke 'kua o Hakipalunuau,
O ka pua na Laka.
O Laka o Hakipalunu, ke kama kuakahi,
Ka poe i moe i ka wai o Punahoa.

I moe i ka wai auanu ka ili,
Ka'u mea 'loha i nalo aku la,
He aloha ko'u e noho aku nei,
He maeele no ka lima.

Gathered at the pool of Punahoa.
You lay in that water,
The pool of rain water that overflows;
Gathered by the 'Āpuakea rain
By the Koholālele wind.
Are you one greeting us?
The rain that breaks on the pandanus
The pandanus from Akiu and
Honokalani;
The pandanus forest of Akiola,
My beloved one who passed away.
My love resides here,
The hand is numb for you,
Beloved chief, beloved chief,
Beloved chief who went away.
The chief was burned,
The flesh was separated,
The chief has taken the form of a spirit.
Many forms; many transfigured forms.
The body of the chief came before god.
The chief became a new deity of Ko'olau.
Returned to the calmness of
Kapueokahi.
The spirits of Kaiakahuli,
The many residing down at the beach of
Nānu'alele.
The chief, the deity of Hakipalunuau,
A descendant of Laka.
Laka of Hakipalunu, the first child.
The people who laid in the water of
Punahoa.
The skin is cold lying in the water.
My beloved one who passed away.
I am dwelling in sorrow,
My hand is benumbed.

(SOURCE: Fornander, vol. 6, 427–429)

For the makaʻāinana, genealogy rooted them to their places. Just by knowing even part of a genealogy, it was often possible to know where a family originated. Because people were so closely connected to their kulāiwi, their family names were often synonymous with their places. I will use the Kūkahiko family of Maui as an example. For generations, the ʻohana Kūkahiko has farmed and fished in Mākena and on the nearby island of Kahoʻolawe, so to those families familiar with Maui, the name Kūkahiko signals Mākena. Similarly, the Nākoa family is rooted in Kahakuloa, Maui. Without hearing the place of residence, a knowledgeable Kanaka hearing the Nākoa family name would likely associate it with the valley of Kahakuloa, just as one might associate the Kūkahiko name with Mākena.

Wahi Pana: The Genealogy of Place Names

Places, like people, have genealogies. Place names serve as historical genealogies, chronicling the changes that have occurred over time in a particular locale. With each passing generation, place names are either passed on to the succeeding generation, forgotten, or renamed. As previously mentioned, the island of Maui, for example, has at least four former names: ʻIhikapalaumaēwa, ʻIhikapulaumaēwa,[22] Kūlua,[23] and Mauiloa.[24] In studying places, it is therefore very important to acknowledge the names bestowed on various places and to understand how and why names are selected and what traditions and histories are attached to each name. To Kānaka and other indigenous peoples who share a close connection to their land and use oral traditions to record their history, place names and landmarks serve as triggers for the memory, mapping the environment and ultimately the tradition and culture of a people.[25] Simply hearing a name can jar a person's memory, taking one virtually back to the place.

Knowing the names of our places is a form of knowledge about our history and heritage. Place names often detail the physical features of the natural environment, enumerate significant historic events of the place, and catalog natural resources of the locale. Those who are unacquainted with the place names of their ancestral lands are ignorant of their own past and the legacies of their ancestors.[26] Place names are of the utmost importance in telling the histories of the landscapes they reference.

Ancestrally, names were given to natural features ranging in size from a small stone to a gigantic mountain on the landscape.[27] Countless names were applied to the landscape. Today, many place names have been forgotten. However, through the use of ancient mele and moʻolelo, some of these place names have been indelibly recorded in Kanaka history.

Moʻolelo: Historical Accounts of Akua

The moʻolelo of akua were also used as a form of performance cartography for ancestral Kānaka, recording the places and deeds of the akua throughout ka pae ʻāina. Kupa o ka ʻāina were caretakers of the moʻolelo attached to their kulāiwi. Moʻolelo were memorized and passed down generation after generation, reinforcing an akua connection and embedding an air of importance to the places enumerated in the moʻolelo.

The island of Maui is named after the demigod Māui; thus, the name bestows much mana upon it.[28] Māui is one of the most prominent demigods in Polynesian mythology, with his exploits extending across the Pacific.[29] Places throughout Polynesia have varying accounts of the moʻolelo associated with Māui and his brothers, but in Hawaiʻi, Māuiakalana is by far the most well-known of the siblings. The four Māui brothers—Māui Mua, Māui Waena, Māui Kiʻikiʻi, and Māuiakamalo—were born to Akalana and Hinaakeahi in the fifteenth wā of the *Kumulipo*.[30]

Māui's parents resided at Makaliua, Maui.[31] Near Makaliua are the famous cliffs of Kahakuloa. According to one version of this story, Māui was the son of Hinalauae and Hina. As an unborn child, Māui snuck out to Kahakuloa to lele kawa (dive off cliffs). Noticing this handsome boy, several fishermen approached him. As they neared, Māui took off running and escaped into the waterfall at Makamakaʻole. There he hid on a dry ledge behind the waterfall until he thought it was safe to resurface. As he fled the waterfall, he was spotted once again. The men pursued him all the way home. Māui ran into the house and into the safety of his mother's womb. When the men peered into the house, the only person present was a pregnant woman beating kapa (bark cloth). The men knew instantly that the child yet to be born was indeed an akua; therefore, they paid tribute to the unborn akua by offering a pig, white chicken, black coconut, red fish, red kapa, and ʻawa (kava).[32]

According to another legend, Kānaka did not have fire until the time of Welaahilaninui. The first fire burnt continuously, fueled by the mana of the gods. Once the gods extinguished it, Māui was determined to learn the secret of fire making. One day, Māui noticed a fire burning in the distance. When he approached, two kupua (supernatural beings) were broiling bananas over the flames. Seeing Māui, the kupua transformed themselves from womanly forms into ʻalae birds (Hawaiian gallinule or mudhen) as they attempted to flee. Having been taught the art of bird catching by his mother Hina, Māui was able to capture the ʻalae named ʻAlaehuapipi. Māui demanded to know the source of fire, threatening to kill the trapped ʻalae if she withheld the secret from him. Fearing Māui would kill her, ʻAlaehuapipi revealed that fire could be obtained by rubbing two sticks of māpele together.[33]

In another moʻolelo, Māui was determined to unite ka pae moku. One day, he went fishing with his brothers at Poʻo, a fishing spot that intersected with landmarks at Kīpahulu and Kaiwiopele in the moku of Hāna. Using his magical fishhook, Mānaiakalani, Māui caught the giant ulua (a type of fish also known as giant kingfish or giant trevally) of Pimoe. Māui warned his brothers not to look back. For two days, he fought with the fish until the brothers ultimately turned back to see what Māui was doing. As they turned, the fishing line broke, releasing the fish. Because the brothers disregarded Māui's command, Māui was unable to unite ka pae moku into a single landmass.[34]

Māui was also famous for slowing the sun for his mother, Hina, who was frustrated that her kapa would not dry before sunset each day. Māui therefore set out to slow the pace of the sun. He traveled to Peʻeloko,[35] Waiheʻe, and there he fashioned a strong coconut sennit to lasso the sun rays. When the sun rose over Haleakalā, Māui snared it with his coconut sennit, then he broke the sun's strongest rays. Convinced that the sun would slow its pace, Māui spared the sun's life. Because of Māui's feat, some say that the true name of the mountain is not Haleakalā (house of the sun) but Alehelā, or ʻAleheakalā (the snaring of the sun).[36]

Māui is also credited with pushing the heavens up into the sky. It is said that Hawaiʻi was filled with darkness because the clouds were so close to earth. In fact, leaves grew flat because they had nowhere else to grow. Slowly, the plants pushed up the sky little by little, enabling people to crawl in the space between the heavens and earth. Concerned for the welfare of the people, Māui instructed a woman to give him a drink of water from a gourd calabash. With renewed strength, standing on the peak of Kaʻuiki in Hāna, Māui pushed the heavens higher, from the level of the treetops, then to the mountaintops, and finally to the spot where the heavens are today.[37] Although clouds may commonly rest upon Haleakalā, they do not rest upon Kaʻuiki, except in heavy downpours, for fear that Māui will thrust them even higher up into the sky.[38] The ʻōlelo noʻeau "Ka ua Lanihaʻahaʻa o Hāna" makes reference to the "rain of the low sky of Hāna."[39]

Moʻolelo such as those about Māui establish the histories and traditions of a place, functioning as a means of place making. Yi-fu Tuan supports this premise by suggesting that "storytelling converts mere objects 'out there' into real presences. Myths have this power to an outstanding degree because they are not just any story but are foundational stories that provide support and glimmers of understanding for the basic institutions of society; at the same time, myths, by weaving in observable features in the landscape (a tree here, a rock there), strengthen a people's bond to place."[40] Moʻolelo record the histories of our kūpuna as legacies for the succeeding generations, evoking a sense of pride in kupa o ka ʻāina, who are able to recite the moʻolelo about akua associated with their kulāiwi.

ʻŌlelo Makuahine: Mapping through Language

Like the many layers of kaona in ʻōlelo Hawaiʻi poetry, place is viewed with immense dimensionality from a Kanaka perspective. Perceptive to the complexity of place, Kānaka use a myriad of terminologies to denote and map place. Moreover, evidence from ʻōlelo Hawaiʻi texts and oral histories suggests that the concepts of place and time were largely indivisible in the ancestral Kanaka worldview.

The four directionals, "aʻe," "iho," "mai," and "aku," represent one tool that Kānaka utilized to denote proximity and location in terms of both space and time. Through the use of these directionals, the elements of space and time are often clarified.

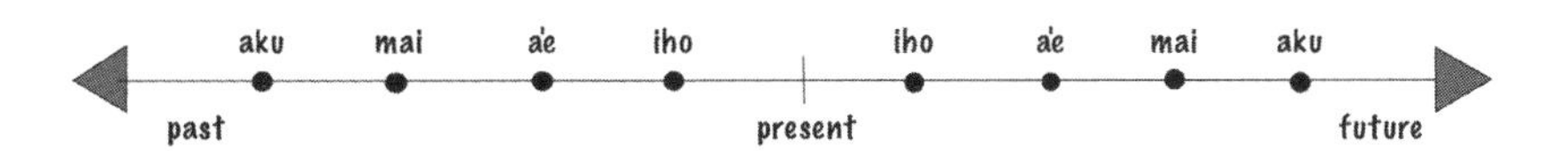

The relationship of directionals to one another in space and time

"A'e" may indicate an upward or sideways movement. It can also suggest that two things are near or adjacent to one another in the sense of place and/or time, as is the case in the following example: "'O ia ko'u mua a'e" (She is the one born just before [or above] me). Another example, "kēia pule a'e" (next week), refers to a time frame adjoining the present. In other cases, "a'e" may be used to indicate a recent past. "A'e" is also used in conjunction with the locative words "luna" and "lalo" to suggest that something is just above or just below something else.

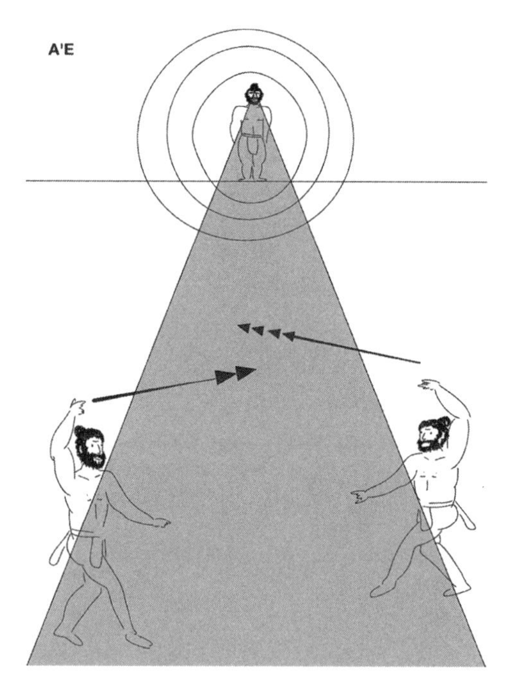

A'e indicates an upward or sideward movement from the perspective of the person speaking

Like "a'e," "iho" can be used to qualify the locatives "luna" and "lalo." "Ma luna iho" is "just above," while "ma lalo iho" translates to "just below." "Iho" implies a downward motion or an action occurring within oneself, such as thinking, eating, or drinking. As a time element, "iho" indicates the near future; thus, it is often translated as "immediately after" and "next" in narrations. Conversely, "iho" can also indicate the recent past, as is the case in "ua 'ai iho nei au" (I just recently ate).

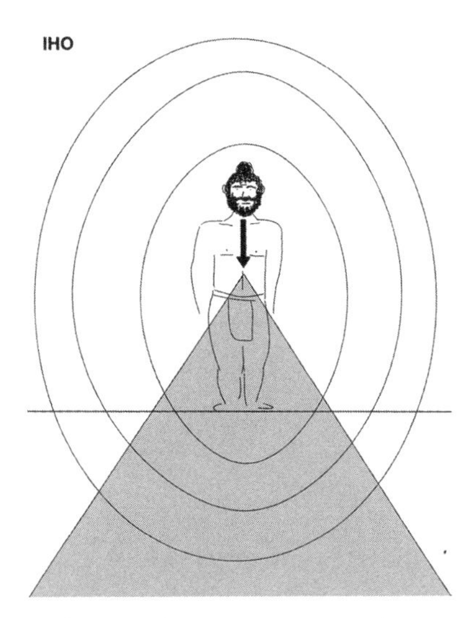

"Iho" indicates a downward action or action within oneself from the perspective of the person speaking

"Mai" often indicates an action toward the speaker and can mean "here" and "this way." "Mai Kapalua mai," for example, means "from Kapalua toward this way" or "this side of Kapalua." "Ma ia wahi mai" similarly means "from that place, on this way."

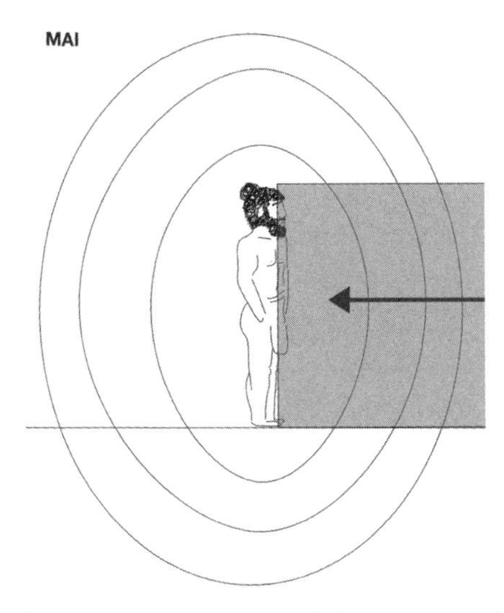

"Mai" indicates an action toward the person speaking

"Aku" contrasts "mai." "Aku" is often used to denote an action that is going away from the speaker or onward, such as "ma Mākena aku" (at Mākena and continuing onward) or "ma ia wahi aku" (from that place farther on). In regard to time, "aku" is often used to denote the distant future or the past, as in "kēlā mahina aku nei" (last month).

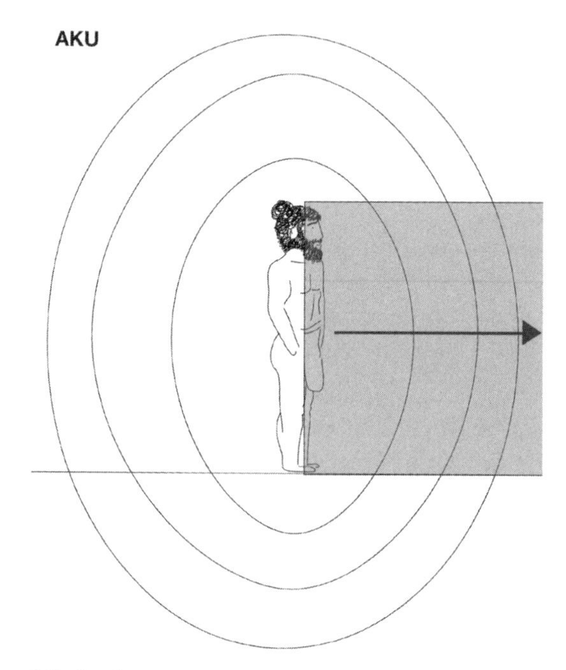

"Aku" indicates an action going away from the person speaking

Another example of how Kānaka used directionals to indicate direction and proximity is the way in which the space over one's head, extending on to the heavens, was named. The solid heavens were known as "ka lani paʻa," "ka lani uli," and "ke ao ulu." Below the solid heavens was the place where the clouds floated, called "luna o ke ao." In descending order, the regions below the "luna o ke ao" were "luna lilo loa,"[41] "luna lilo aku," "luna loa aku," and "luna aku." "Luna aʻe" was the region directly over one's head when standing.[42] People stood on the surface of the earth, known as "lalo." Below "lalo," descending farther below the earth, were "lalo o ka lepo," "lalo liloa" (an abbreviated form of "lalo lilo loa"), and "lalo ka papa kū."[43]

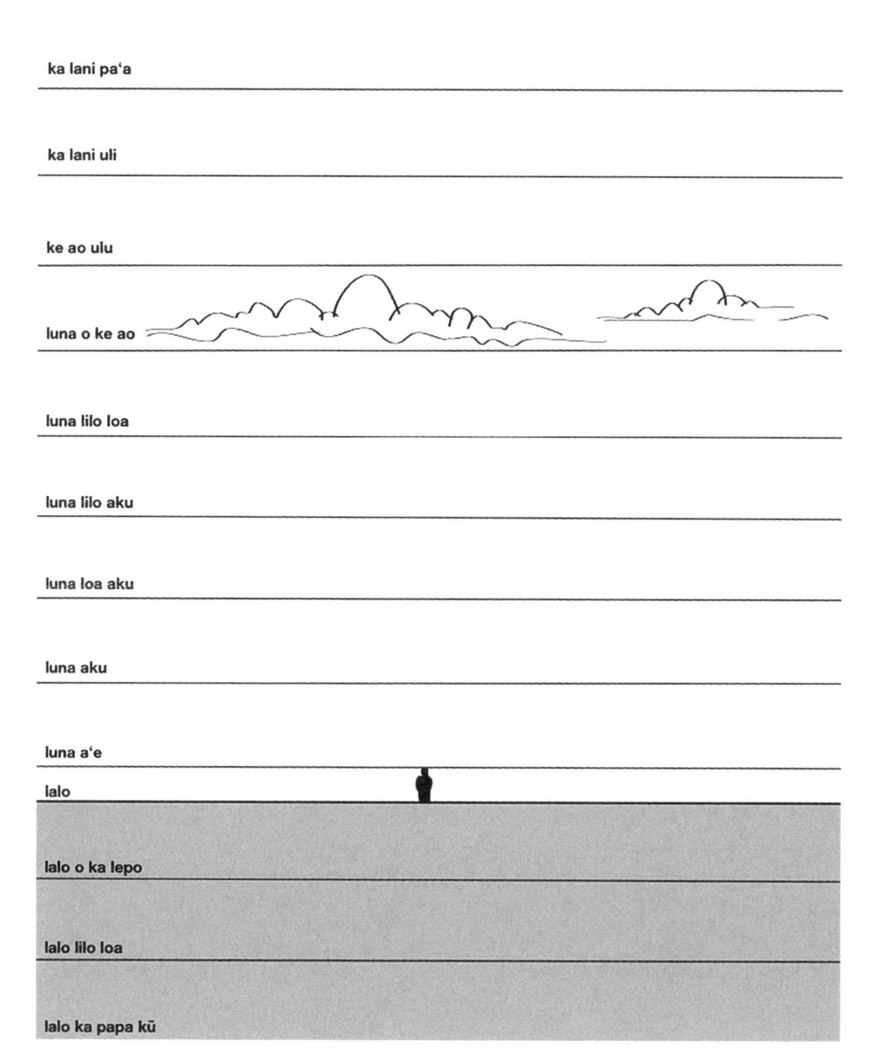

The strata of the skies and the earth

While the use of directionals to indicate actions away from, toward, above, and below a speaker may be common in many languages, perhaps the most universal way of mapping one's environment is through the use of the four kūkulu (cardinal points).[44] North, south, east, and west are, respectively, "'ākau,"[45] "hema,"[46] "hikina,"[47] and "komohana"[48] in 'ōlelo Hawai'i.[49]

The use of locatives also allows Kānaka to map our world through the intricate use of our language.[50] ʻŌlelo makuahine locatives, like those of some other Oceanic languages, are treated differently than they are in English. Locatives such as "luna" (above), "lalo" (below), "waena" (middle), "ʻaneʻi" (here), "ʻō" (there), "waho" (outside), "mua" (front), "hope" (back), and "loko" (inside) function like proper nouns indicating places when they serve as subjects or direct objects of a sentence.[51] Here are a few examples:

Nani ʻo Maui.	Maui is beautiful.
Nani ʻo waho.	Outside is beautiful.
ʻIke au iā Maui.	I see Maui.
ʻIke au iā waho.	I see the outside.

From a Kanaka perspective, then, "waho" is fundamentally used in the same way that place names such as "Maui" or "Hawaiʻi" are. In this way, Kānaka are able to map places by either referring to the place name or the place's location on the heavenscapes, landscapes, or oceanscapes.

Another method of mapping place and time is through the use of the proximity markers "nei" and "lā."[52] "Nei" suggests close proximity, while "lā" implies a greater distance, as is illustrated below:

nei	here and now, happening now in this time and place
lā	there, denotes distance in place
ʻo ia nei	(s)he near
ʻo ia ala	(s)he far
ko ia nei	his/hers near
ko ia ala	his/hers far
ke hele nei ʻo ia	(s)he is currently going (happening in close proximity to speaker)
ke hele lā ʻo ia	(s)he is currently going (happening far from speaker)
ua kāne nei	this aforementioned man
ua kāne lā	that aforementioned man

Notice in the examples above, "nei" indicates closer proximity to the speaker in both time and space than does "lā."[53]

Other ʻōlelo Hawaiʻi examples of mapping place through the use of language are the determiners "kēia" (this), "kēnā" (that, near the person being spoken to), and "kēlā" (that, far away from the speaker and person being spoken to). What differentiates these determiners from one another is the proximity that they connote.

By observing clues embedded in ʻōlelo Hawaiʻi, we are able to gain insight about how ancestral Kānaka constructed and mapped their places orally. The aforementioned examples by no means comprise an exhaustive list of terminologies denoting location, directionality, and proximity; they merely highlight some of the ways in which Kānaka used their mother tongue in creating performance cartographies to map their places.

ʻŌlelo Noʻeau: An Intimate Understanding of Place

Nearly 40 percent, or 1,149, of the 2,942 proverbs cited in *ʻŌlelo Noʻeau* relate to places, illustrating how important the sense of place was in ancestral times.[54] ʻŌlelo noʻeau tied to different locales shed light about those places. Maui, for instance, was referred to as "ka mokupuni kuapuʻu" (the hunchbacked island) because its shape resembles the figure of a hunchbacked person (ON 161). The parameters of Hāna were given in the saying "Hāna, mai Koʻolau a Kaupō" (Hāna, from Koʻolau to Kaupō) (ON 55). The boundaries of Waiʻehu were similarly defined as extending from the cliff of Kapulehua to the cliff of ʻAʻalaloa in the ʻōlelo noʻeau "Waiʻehu mai ka pali ʻo Kapulehua a ka pali ʻo ʻAʻalaloa" (ON 318).

Maui, Ka Mokupuni Kuapuʻu

Places were often associated with ali'i in 'ōlelo no'eau, thereby mapping the island through the use of the names of famous ali'i as well as the characteristics of the places the ali'i frequented. "Ka pali kāohi kumu ali'i o 'Īao," for instance, poetically references "the cliff that embraces the chiefly sources of 'Īao" (ON 165). 'Īao was a significant place for ali'i nui. Many of the highest-ranking ali'i throughout ka pae moku were buried there, making it an extremely sacred place. In fact, 'Īao is one of the most sacred places on Maui. "Ka malu ao o nā pali kapu o Kaka'e" (the shady cloud of the sacred cliffs of Kaka'e) (ON 159) and "Ka pela kapu o Kaka'e" (Kaka'e's sacred flesh) (ON 166) are two other poetic references to 'Īao, glorifying it as the final resting place of Kaka'e, an ancient ali'i nui of Maui.

Many 'ōlelo no'eau map the landscape by describing natural phenomena characteristic of particular places. "Ka nalu he'e o Pu'uhele" boasts of the good surf at Pu'uhele, Hāna, Maui (ON 162), while "ka pali hinahina o Kā'anapali" draws attention to the gray hills of Kā'anapali (ON 165). Even without having been to Kahului, Maui, a person hearing the 'ōlelo no'eau "ke kai holu o Kahului" would know that Kahului is famous for its swaying sea (ON 185).

The interconnectedness of ancestral Kānaka and the places they called home is apparent in many 'ōlelo no'eau. Such proverbs interchange the names of the places for the people themselves. Some people hearing "Kaihalulu i ke alo o Ka'uiki" (Kaihalulu lies in the presence of Ka'uiki) for the first time might assume that this 'ōlelo no'eau simply refers to Kaihalulu and Ka'uiki, two places in Hāna, Maui, that are in close proximity to one another (ON 151). However, the place names honored in this 'ōlelo no'eau also refer to the people of Kaihalulu and Ka'uiki, who are often found in the company of one another. "Nāhiku hauwala'au" suggests that the people of Nāhiku talk very loudly (ON 242). "Honua'ula kua la'ola'o" means "callous-backed Honua'ula," referring to the hardworking people of Honua'ula, who develop calluses on their backs from carrying heavy loads (ON 113). The diligent people of Kula are praised for accomplishing their goals in the saying, "nā keiki uneune māmane o Kula" (the children of Kula, who tug and pull up the māmane) (ON 245).

Many place-centered 'ōlelo no'eau recount famous events of that area. "Ho'i hou ka wai i uka o Ao" (the water returns to the upland of Ao) originated during the battle of Ka'ua'upali in 'Īao fought between Maui forces and those of Kamehameha (ON 110). The battle was so fierce that the stream was dammed with the numerous bodies of the slain warriors. This 'ōlelo no'eau recounts how people had no option but to travel deep into the valley to find uncontaminated water for household use following the battle. "Ka wai hoʻihoʻi lāʻī o ʻEleile" (the water that returns the ti leaf of ʻEleile) makes reference to ti stalks thrown into the famous pool, ʻEleile (ON 178). Observers delighted in seeing the ti stalks being carried to and fro in the pond before finally flowing downstream. "Pākahi ka nehu a Kapiʻioho" (the nehu of Kapiʻioho are rationed, one to a person) refers to times when fish are scarce and must be carefully rationed (ON 284).

These (and all) ʻōlelo noʻeau provide insights into an ancestral Kanaka worldview. While Mary Kawena Pukui's book, *ʻŌlelo Noʻeau,* contains the most comprehensive listing of ʻōlelo Hawaiʻi wise sayings to date, her book does not exhaust all known ʻōlelo noʻeau. Nevertheless, the fact that so many of the ʻōlelo noʻeau compiled in her book are directly related to specific places reveals the undeniable connection that Kānaka enjoyed with their places. ʻŌlelo noʻeau grounds ancestral Kanaka identities and maps their places.

Makani and Ua: Mapping the Elements of Nature

Kānaka mapped by naming individual winds and rains of their places.[55] Winds and rains were place specific; their nature differed from place to place, as did their names. The soft, gentle Līlīlehua rain of Kāʻanapali did not, for example, have the same name as Hāna's misty white rain, Kauakea. Even within a particular locale, there might be several distinct winds and rains, each having its own characteristics and names.

Like place names, winds and rains acted as mnemonic devices facilitating the recollection of the places they occurred. Moʻolelo, mele, and ʻōlelo noʻeau often enumerated the winds and rains of various places, describing their individual traits.[56] By listening to a moʻolelo, mele, or ʻōlelo noʻeau or witnessing a hula, people could be virtually transported back to the places with which the winds and rains were associated. The Moaʻe wind of Honuaʻula was known for its force. People would therefore say, "Honuaʻula, e pāluku ʻia ana nā kihi poʻohiwi e nā ʻale o ka Moaʻe," in poetic reference to a person battered by the wind of Honuaʻula (ON 113). Another gusty wind was that of Waikapū, "ka makani kokololio o Waikapū" (ON 159). The Maʻaʻa wind of Lele, known today as Lahaina, was characterized as a coconut leaf lifting wind in the saying, "ka Maʻaʻa wehe lau niu o Lele" (ON 157). Pāʻia was known for its dust-smearing wind, "ka makani hāpala lepo o Pāʻia" (ON 158).

Some ʻōlelo noʻeau did not merely elucidate the nature of the winds and rains but also identified specific phenomena associated with them. "Ka makani kā ʻAhaʻaha laʻi o Niua" alludes to "the gentle ʻAhaʻaha breeze of Niua that drives in the ʻahaʻaha fish" (ON 158). In this example, ʻAhaʻaha refers to both the fish and breeze of the same name. Fishermen knew that when this breeze blew, it was the right time to launch their canoes in search of the ʻahaʻaha fish. The Hau wind of Kula was noted for its ability to blow smoke swiftly; a person who was swift was likened to the Hau wind.

Winds and rains also facilitated in mapping directions. Kona winds, regardless of island, blew from the south. Even a person unfamiliar with a place would know what direction south was, should a kamaʻāina say that the Kona winds were blowing. By knowing where south was, the other three cardinal points could be likewise referenced.

Mele a Kūapākaʻa

The mele that follows maps various winds and rains on the island of Maui. It is an excellent example of how "local knowledge" is embedded in Kanaka performative cartographic practices. This mele appears in a story related to Pākaʻa and his son, Kūapākaʻa.[57] In this moʻolelo, Pākaʻa is the favorite caretaker of the aliʻi Keawenuiaʻumi of Hawaiʻi Island. Out of appreciation for the outstanding care that Keawenuiaʻumi receives from Pākaʻa, Keawenuiaʻumi rewards Pākaʻa with many lands. However, two other caretakers are jealous of Pākaʻa's close ties to the aliʻi and conspire against him, boasting of their own abilities and degrading those of Pākaʻa. Impressed by the skills of these caretakers, Keawenuiaʻumi relinquishes Pākaʻa's landholdings and favors Hoʻokeleihilo and Hoʻokeleipuna. It does not take long for Keawenuiaʻumi to realize he has made a grave mistake. His new right-hand men do not compare in the least with Pākaʻa. Keawenuiaʻumi therefore orders his servants to search for Pākaʻa, his beloved caretaker.

In the meantime, Pākaʻa teaches his son, Kūapākaʻa, all of the skills necessary to be a great caretaker. Certain that Keawenuiaʻumi will search for him once he realizes the faults of his new caretakers, Pākaʻa develops an elaborate plan to win back his lands and the praises of Keawenuiaʻumi. He prepares everything necessary to host the aliʻi and his party. When the party of the aliʻi approaches, Kūapākaʻa warns the aliʻi and his kāhuna that it will be a stormy day and that it is not wise to sail. The kāhuna point out to Kūapākaʻa that it is a calm and sunny day and that no harm will come to them. In order to demonstrate his knowledge, Kūapāka'a recites chants revealing the winds and rains on all of the islands. Finally, when the party dismisses his warning and sets out to sea, Pāka'a orders his son to open the magic wind gourd and use the winds and rains against the party. Upset that his kāhuna did not foresee that a storm would hit, Keawenuia'umi is outraged. The group is forced to return to land as Kūapāka'a had previously suggested. Kūapāka'a and Pāka'a ultimately win back the favor of Keawenuia'umi.[58]

Together this mo'olelo and the mele that accompany it epitomize performative cartography. Through the mo'olelo, the foundation for the mele is laid. The mele itself reflects the depth of knowledge that Kānaka had about their places. Each place had its own winds and rains—all were named. The name of each wind and rain revealed its nature; some were gently blowing winds, while others were destructive rains. Through mele such as this, the names, physical locations on the landscape, and characteristics of winds and rains were mapped.

Kiauau! Kiauau!! Kiauau!!!
Aia la o ka pali ale ko Hilo Waiakea
makani,
He Aimaunu ko Hana,
He Ailoli ko Kaupo,
He Moae ko Kahikinui,
He Papa ko Honuaula,
He naulu[59] ae i Kanaloa,
Hina ka hau[60] i ka uka o Kula,
Ke noke mai la i ke pili,
Ka makani o Kula o ka Nau,
Ulalena i Piiholo,
Ukiu ko Makawao,
Ka ua Elehei i Lilikoi,[61]
Ka ua Puukoa i Kokomo,
Ka Haule aku i Mauoni,
Ka Hau aku i Kealia,
He Kaumuku ko Papawai,
Olaukoa i Ukumehama,[62]
Makani wawahi hale i Olowalu,
Kilihau iho no ilaila,
Kololio mai o Waikapu,

Gently! Gently!! Gently!!!
Here is the pali 'ale wind of Hilo,
Waiākea,
The 'Aimaunu belongs to Hāna,
The 'Ailoli of Kaupō,
The Moa'e of Kahikinui,
The Papa of Honua'ula,
The nāulu of Kanaloa,
The hau blows to the uplands of Kula,
Seeking the associate,
The wind of Kula is the Nau,
The 'Ulalena in Pi'iholo,
The 'Ūkiu of Makawao,
The 'Elehe'i rain in Liliko'i,
The Pu'ukoa rain in Kokomo,
The Hā'ule is at Mauoni,
The Hau is at Keālia,
The Kaumuku of Papawai,
The 'Ōlaukoa in Ukumehama,
The house destroying wind in Olowalu,
The Kilihau is also there,
The Kololio of Waikapū,

Ka Iaiki[63] ko Wailuku,	The I'aiki of Wailuku,
Ka Oopu ko Waihee,	The 'O'opu of Waihe'e,
Pa ka makani Kauaula,	The Kaua'ula wind blows,
Ke nu mai la i na pali,	It roars along the cliffs,
I na pali aku o Kahakuloa,	On the cliffs of Kahakuloa,
O Waiuli aku i Honolua,	The Waiuli is next at Honolua,
Pohakea i Mahinahina,	The Pohākea is at Māhinahina,
Lililehua i na pali,	The Līlīlehua is at nā pali,
Kaimihau ko Kekaa,	The Kaimihau of Keka'a,
Nahua i Kaanapali,	The Nahua at Kā'anapali,
Ka Ululoa i kela pea,	The Ululoa is that border,
Ka Ma-a-a[64] ko Lahaina,	The Ma'a'a of Lahaina,
Ke kau mai la i Kamaiki,	The Kamaiki perches,
Moaeae aku la ka pali,	The Moa'ea'e along the cliff,
Ka Alani ko Liloa,	The Ālani of Liloa,
Ka Paalaa[65] o na Kaha,	The Pa'ala'a of na Kaha,
O na keiki a Ku, a Naiwi,	The children of Kū, of Nāiwi,
Kaiaulu[66] i Pulupulu,	The Kaiāulu is at Pulupulu,
Ke holio mai la i ke kula,	The wind that doubles up on the plains,
Holokaomi Paomai,	The Holokaomi of Paoma'i,
He Kupa he Okoe [sic] ka makani,	The Kupa the 'Ōke'e wind,
He Pelu ka makani o kai,	The Pelu wind of the lowlands,
Paiolua i ka moana,	The Paiolua is on the ocean,
Ka Moae, o ka Hoolua	The Moa'e, the Ho'olua

(SOURCE: J. H. Kanepuu, "Kaahele ma Molokai: Helu 5," *Ke Au Okoa,* October 17, 1867.)

Conclusion

The fact that Kānaka had a very close connection to the 'āina in ancestral times is evident in our 'ōlelo makuahine. Terms such as 'āina,[67] aloha 'āina (love for the land), and kua'āina (the people who carry the burden of the land on their backs) all reflect an undeniable bond between the 'āina and Kānaka. Kānaka knew their places so intimately that they were able to describe in great detail the characteristics that set their kulāiwi apart from other places. With pride, Kānaka utilized performance cartographies to map their places and to record their legacies. Through eloquent mele, gracious hula, and captivating moʻolelo, talented Kanaka artisans kept people and places alive via their compositions by commemorating the histories and natural features of their ancestral places and by acknowledging the accomplishments and stellar characteristics of their ancestors.

Chapter 5 Ancestral Sense Abilities

Kānaka have been able to navigate throughout the Pacific for thousands of years by sensing subtle changes in our environment. We navigate by observing the stars, sun, moon, clouds, ocean currents, birds, and sea life; listening to the rhythm of the ocean as it hits the sides of the canoe; and feeling the way the canoe glides. Navigation requires us to tap into our abilities to see, hear, touch, and smell our surroundings. The use of our senses (ʻike), combined with thousands of years of ancestral experience, enables us to sail to distant lands purposefully and intentionally.

This chapter explores the ways in which Kānaka developed keen intellectual perceptions informed by our interactions with our environment and our kūpuna. Our deep consciousness and appreciation for the natural environment are reflected in the many "sense abilities" that Kānaka have refined over time. In this context, a "sense ability" is the capacity to receive and perceive stimuli from our oceanscapes, landscapes, and heavenscapes and to respond to these sensory stimuli in ways that contribute to our overall understanding of our world. It is an awareness that our environment is constantly sending us information, but unless we hone our abilities to sense the world around us, much of this information can go undetected and unappreciated. This chapter demonstrates that our sense abilities contribute to our human experiences and serve as the foundation of Kanaka knowledge systems as well as Kanaka geographies.

The Sense Ability of Sight

Knowing and seeing is one and the same in ʻōlelo Hawaiʻi: ʻike. The word "ʻike" is further defined in the *Hawaiian Dictionary* as "to feel, greet, recognize, perceive, experience, be aware and understand."[1] These multiple meanings are illustrative of a worldview that relies on the senses to gain insight. Ancestral Kānaka were so in touch with their natural surroundings that they were able to ʻike maka (observe and identify with one's own eyes) even slight changes in weather patterns. By utilizing this visual knowledge, they accurately predicted when many natural phenomena would occur. From cloud formations signaling landfall to a navigator to massive schools of ʻāweoweo fish indicating the death of an aliʻi, these types of visual clues continue to be instrumental for Kānaka in comprehending our environment.[2]

The fact that many generations of Kānaka valued the quest for deep knowledge and insight known as "ʻike kūhohonu" is reflected in our ʻōlelo makuahine and our worldview. Words such as "ʻikeʻike" (a reduplication of the word ʻike), "ʻike loa" (to know extensively, to be well versed), and "ʻike kumu" (fundamental knowledge) all have the connotations of both seeing and knowing the world through a Kanaka lens. Only a people intimately connected to nature are able to see the correlation between scenery (ʻikena) and knowledge (ʻike), coining a single defining word, "ʻikena," to refer to both concepts. The expression "ʻaʻohe kio pōhaku nalo i ke alo pali" (on the side of a cliff, not one jutting rock is hidden from sight) even suggests that there is no use in concealing anything because the observant eye sees all.[3]

Close observation of the elements around us is crucial to our existence. To know something well—to have certain knowledge—requires us to ʻike pono (to see and observe well). Through this process of seeking ao (enlightenment) via hākilo pono (close observation), we become an aokanaka (enlightened person).

A hallmark of an aokanaka is the ability to recognize hōʻailona not commonly noticed by the untrained eye. Akua sometimes give people the gift to see what others cannot. Those blessed with the rare gift of ʻike pāpālua (double knowledge or the ability to see the future) are able to interpret what the supernatural world is communicating to them through their extrasensory perception. In ancestral times, chiefs sought the advice of highly revered kāula (prophets), who possessed the gift of reading the hōʻailona in nature. Through contextual clues, kāula and others sensitive to the signs laid out for them by their environment, akua, and ʻaumākua predicted the future and planned appropriate courses of action. Kāula were the "noiʻi i nā mea ʻike ʻole ʻia" (seekers of unseen things).[4]

Hōʻailona are still relevant in a modern context. Seeing a white pueo (owl) flying over a highway while driving may be meaningless to some, but others may see it as a hōʻailona to slow down, lest the driver get into an accident. Likewise, a double rainbow may be no more than a beautiful sight to some, but for others attending a double funeral it may symbolize the spiritual return of two teenagers killed in a tragic accident. Similarly as a haumāna (student) of a hālau hula (school of hula), I often go into the forest with my hālau to gather foliage for our lei. Whenever we go, we always begin with an oli (chant) to request permission to enter the forest. Once the oli is completed, my kumu hula observes the forest in search of hōʻailona indicating whether or not the forest will grant us permission to enter. The hōʻailona is often subtle, sometimes appearing as a bird landing nearby or as a gentle breeze blowing through the trees.

Hōʻailona manifest themselves in an array of ways. Sometimes the mere sight of an object or place may unexpectedly conjure up haliʻa (fond recollections of loved ones) of a deceased loved one; yet hōʻailona are more than just memories. They are signs or omens that may appear as visions or premonitions of various forms, including but not limited to moe ʻuhane (dreams), hōʻupuʻupu (horrifying visions about someone else), kupu wale (spontaneous dreams), noʻonoʻo mua (dreams from previous experiences), moe hoʻokō (visions that later come true, fulfilling a prophecy), and moe ʻawahua (bitter dreams).[5] Visions may be seen during the day as hōʻailona or at night as hōʻike a ka pō (night visions), appearing when one is either awake or sleeping. Sometimes information is revealed to the seer in the form of an akakū (trance) or a vision resembling a reflection on the water. Haili moe appear in dreams from the darkness, as do hihiʻo. Moe piʻi pololei (prophetic dreams) provide insight about what is to happen in the future. Inoa pō (names to be given to children appearing in dreams) are a form of hōʻailona that often comes in the dreams of close friends or family members.[6]

Sometimes hōʻailona are not abstract images understood only by a few; rather they are symbols of cultural significance readily understood by the general population. For example, the color red was sacred to aliʻi; the sight of a canoe with a red sail alerted all on shore that an aliʻi would soon be landing and that proper protocol was required. Some kūpuna still refuse to wear a red malo (loincloth) during protocol ceremonies, stating that the color red is reserved for ali'i and inappropriate for use by maka'āinana.

The colors worn by the maka'āinana were hō'ailona as well, signifying where a person was from and from whom they descended. Because the hues varied from region to region, the color of one's kapa was often associated with one's homeland. The motifs printed on kapa likewise marked a person, as some designs were reserved strictly for a particular family; only someone of that genealogical line could wear the design.[7]

Even today, in preparation for hula performances and competitions, kumu hula must carefully select the style and color of the costumes that their dancers wear. Performing a hula in honor of Pele (goddess of the volcano) while wearing red, a color associated with lava, would be expected and appropriate, while a purple costume would likely receive critical reviews from the judges.

Over many generations, Kānaka have developed a keen understanding of our world through careful observation of hō'ailona and subtle physical changes in the environment. As heirs to the 'ike kupuna (ancestral knowledge) of those who came before us, we understand that the rising and setting of stars is linked to the season of the year. We recognize that certain nights of the month are better suited for preparing fishing and farming implements rather than actually fishing and farming. And, we know that many natural phenomena may be accurately predicted if we are able to see the hō'ailona in the environment. As we strengthen our sense ability of sight, we gain insight from our environment and add to the knowledge base of our kūpuna.

The Sense Ability of Listening

In ancestral times, the sound resonating from a bell stone may have signaled danger, while a conch shell may have announced the arrival of a sacred ali'i. The sound of women beating kapa was encoded with hidden meaning; women often used their i'e kuku (anvils) to communicate with others within hearing range. The sound of kalo being pounded into poi or kapa being beaten indicated that the sun had not yet set in the direction that the sound was coming from, as it was generally feared that loud sounds at night would arouse and summon spirits.[8] Our aural sense is an important tool for knowledge acquisition and comprehending the world.

The sounds of fish thrashing, branches swaying, water rushing, trees falling, and dogs barking all provide audible clues about our surroundings as a

Kanaka geography. Taking the time to be silent and attuned with the music of nature not only allows us to more actively listen to the environment but also to listen to ourselves and our kūpuna. When we focus internally, information hidden just below the surface of the conscious mind is sometimes exposed in the form of soft inspiration. On occasion, ʻūlāleo (unusual supernatural sounds) are heard. Such sounds or voices are often believed to be the akua speaking to us. The quieter we are, the more apt we are to hear. It is not surprising, then, that even when people are terminally ill and can do little more than lie in bed, they can often still hear what is going on around them. Hearing is usually the last of the five major senses to go.[9]

Yet hearing is not always the same as listening. The more we utilize our sense abilities, the more acute they become. Most people are born with the sense of hearing, but developing one's sense ability as an active listener takes practice; however, over time, with practice, active listening can become second nature. When a person is an active listener and is attuned to one's surroundings, it is often possible to listen to one's environment effortlessly.

I vividly remember the moment in which I realized for the first time that my own sense ability of listening transitioned from hearing to listening. I was living in seclusion on my kulāiwi in a remote valley in order to focus on completing my doctoral dissertation. One day as I was writing intently on the porch of my family's home, I heard a sound that piqued my interest. I instinctively knew that a small flash flood had just occurred. Despite being a hundred yards away from the stream and being engrossed in my dissertation writing, I recognized the subtle change in the sound of my environment. I walked to the stream to find that the water level had indeed risen a few inches. In that instant, I understood how critical the sense ability of listening was to ancestral Kānaka who relied on their environment for sustenance.

While listening to the sounds of nature is informative, the spoken word is perhaps the most important of the aural knowledge bases for Kānaka. Ancestrally, ʻōlelo Hawaiʻi was an oral language. It was therefore imperative that keiki (children) listened carefully to the teachings of kūpuna and other masters, soaking up and internalizing what was shared with them. Many Kānaka continue to pass on our knowledge orally, making it important that we listen to the moʻolelo and moʻokūʻauhau that our kūpuna share with us.

This sentiment is reflected in the word "mānaleo." Derived from two words—"māna," referring to food that has been chewed by a parent for a child, and "leo," voice—"mānaleo" is a metaphoric reference to how the voice of native ʻōlelo Hawaiʻi speakers is fed to succeeding generations. The term was coined in the late 1970s during the Hawaiian Renaissance, at a time when Kanaka scholars acknowledged that the number of people who spoke ʻōlelo Hawaiʻi as their first language was rapidly declining. Thus, scholars began recording, broadcasting, and archiving oral histories of mānaleo. By accessing and listening to mānaleo audio and video recordings preserved in archival collections, we learn more about the intricacies of the ʻōlelo makuahine of ka pae moku and gain insight about the worldviews of native speakers.

Like mānaleo recordings, a vast legacy of oral traditions has also been preserved via musical recordings. From the physical features of various landscapes to the multitude of speech patterns enumerated in song, mele continue to inform current generations of Kānaka about our history and our forms of poetry. Although only a few hundred mānaleo remain, mele written in ʻōlelo Hawaiʻi are serving as a resource for Kānaka eager to learn about the traditions of our kūpuna. The ʻike kupuna contained in archival collections is priceless because these collections capture the worldview of these mānaleo by quoting them through their lyrics. These fragments of ancestral customs and practices are treasures for this generation as well as those to come in the future.

While we are fortunate that these archival collections are accessible today, it is equally important for current generations to show an interest in the ʻike of today's kūpuna. Like their ancestors, many kūpuna today are in search of people worthy of being taught the knowledge they possess; however, displaying an interest in an ancestral practice does not guarantee that you will be selected as a haumāna. Masters of trades do not always have an open-door policy that allows anyone and everyone to study with them. In fact, some kūpuna handpick their haumāna based on their own undisclosed criteria. Only those who the masters feel are worthy of such knowledge are given the opportunity to learn. Should these masters be unable to locate worthy haumāna, their ʻike will likely pass on with them and possibly be lost forever. It is important to Kanaka knowledge systems that we seek out and listen to respected cultural practitioners if we are to maintain our unique cultural identity as Kānaka.

Staunch practitioners of Kanaka traditions often prefer understudies who are keen observers, as opposed to those who are verbally inquisitive, because it is often believed that many of the answers to the questions that we ask are right before our eyes. When we focus too much on asking questions, we fail to notice the hōʻailona before us; when we are observant, and the time is right, answers often present themselves. Understudies are therefore encouraged to "pa'a ka waha, hana ka lima" (close our mouths, work with our hands) (ON 281).

The Sense Ability of Taste

The sense ability of taste is grounded in the places where we are raised; thus, the sense ability of taste is also a form of Kanaka geography. While it may be remarkable that some people's taste buds are so acute that they are able to differentiate the tastes of varieties of bananas or species of fish, what is even more amazing is that some Kānaka attest that they are able to taste the terroir or the regional differences in foods. Different types of soil, climate, and water conditions produce kalo with a taste distinct from that grown in other areas.

Just as the terroir of foods may instantaneously conjure up thoughts of the places in which those foods were grown and produced, when we indulge in the delicacies of our kulāiwi, our sense of taste may unexpectedly and instantaneously facilitate in recalling events of the past. The taste of foods can jar the memory, taking people back not just to the places but also times and events that had previously been suppressed in one's mind. I remember biting into a mountain apple and flashing back to a time when I once sat under the mountain apple tree in my grandparents' yard, eating the fruits to my heart's delight.

The fact that people living in different regions of the world have palates that are informed by their regional preferences reinforces the notion that the sense ability of taste is linked to where one is born and raised. We tend to acquire a taste for the foods that are readily available in our kulāiwi. In this way, the sense ability of taste is also a sense ability of place. Our sense ability of taste is a reflection of who we are and where we come from. It is no wonder then that when we eat foods, we sometimes think about the places with which we associate those foods.

The Sense Ability of Touch

Many ancestral Kanaka practices were tactile and encoded with meaning. Recognizing the textures of the various stages of kapa making was necessary in knowing whether kapa was adequately and properly beaten. The touch and feel of fine makaloa (a native sedge) mats were signs of prestige and status; the finest mats were reserved for the ali'i. For an ancestral healer, touching someone's body furnished clues about that person and how they carried their stress.

Our ability to feel the land and sea is a source of Kanaka knowledge, as we have fostered an intimate relationship with the 'āina over many generations. By being rooted in our kulāiwi, we not only connect with nature but also reaffirm our relationship with our kūpuna who once walked the very same lands. As we feel the water flowing into a lo'i to ensure that it is at the optimal temperature for kalo to thrive, or as we feel and identify the various winds and rains unique to our kulāiwi, we walk in the footsteps of our ancestors. We honor and perpetuate our ancestral knowledge. We engage in a Kanaka geography by exercising our sense ability of touch.

The Sense Ability of Smell: Scents of Place

Kānaka often rely on our sense ability of smell as a knowledge base for comprehending our surroundings as well. Without the sense ability of smell to immediately alert us of danger, a small fire ignited by dry grass could develop rapidly into a large brush fire, threatening life and property. Savory food, fragrant flowers, and decomposing rodents are all smells that assist us in connecting with our environment.

Sometimes smells not only connect us to our environment, but, like other sense abilities, they can also jar the memory and connect us to previous experiences, times, and places. Even years after his death, the smell of the cologne that my grandfather once wore transports me back to my childhood, recalling the fond memories that he and I shared. Our sense abilities are a means by which we glimpse back in time to better understand our kūpuna. Utilizing ancestral methods of understanding our surroundings assists us in tapping into our 'ike kupuna. Like a baby who instinctively knows the scent of its mother, Kanaka scholars have the potential to instinctively sense the ancestral and geographical knowledge of our kūpuna. When we are quiet and allow ourselves to sense our surroundings, we become more intimate with our kūpuna and the places they called home.

The Sense Ability of Na'au

On a daily basis, our experiences are enriched by our sense abilities of sight, hearing, taste, touch, and smell. Yet these five sense abilities make up only a part of our experiences. The sixth sense ability is often associated with one's intuition and supernatural phenomena that defy "rational" logic. For Kānaka, this sixth sense ability emanates from our na'au as both 'ike kumu and 'ike hānau (a knowledge base with which Kānaka are born).

Lorrin Andrews defines "na'au" in *A Dictionary of the Hawaiian Language* as "the small intestines of men or animals, which the Hawaiians suppose to be the seat of thought, of intellect and the affections."[10] In their *Hawaiian Dictionary,* Pukui and Elbert have a similar definition: "intestines, bowels, guts; mind, heart, affections of the heart or mind; mood, temper, feelings."[11] While the na'au is one's innards, it is also the center from which one's instincts and feelings radiate.[12]

Our gut feelings provide us with insight for future courses of action. Unlike hō'ailona, which are signs or signals from external sources, na'au is made up of feelings or gut reactions that emanate internally. In times of distress or fear, a Kanaka might say, "I feel like my na'au is turned inside out," or "My na'au is telling me something is not right." The na'au can therefore affect the way a person approaches a given situation. A mother, sensing in her na'au that harm will come to her child at the beach, would likely listen to her 'ike hānau, forbidding her son from swimming, fishing, surfing, or canoe paddling that day. Others insist that they know the exact time that a close relative or friend miles away died because of a feeling in their na'au. Sensing that one's child is in danger or that a loved one has died are messages that some Kānaka feel intuitively, and it makes us know wholeheartedly that our na'au is speaking to us.

Like an umbilical cord that connects mother to child, the na'au is a spiritual link between ancestor and descendant. It is from the na'au that Kanaka ancestral knowledge emanates, strengthening the bond with one's ancestors, those both known and unknown. Today, highly skilled Kānaka, following in the footsteps of their ancestors, practicing the same trades as their kūpuna, are sometimes said to excel because "it is in their genes." That is to say, the ancestral knowledge of their kūpuna runs through their veins and na'au.

By fully utilizing all of one's sense abilities, people are able to have an intimate connection with their environment. For some Kānaka, when their bond with nature is weakened, their body suffers various ailments. After one of my mother's lomilomi (massage) sessions, she advised a female Kanaka client to take time to bond with nature, walking barefoot on dirt, and spending more time in her yard.

One's naʻau affects the rest of the body so profoundly that it is often equated with one's character. "Naʻau" therefore serves as a root word for many other terms. Naʻauao, for example, is a person who is ao (enlightened). Conversely, someone who is naʻaupō, one whose intestines are dark, is said to be ignorant and left in the dark. Benevolent people are filled with naʻau ahonui, naʻau aliʻi, naʻau palupalu, and naʻau aloha, while those who are evil or nasty have naʻau ʻinoʻino, naʻau keʻemoa, and naʻau kopekope. Righteousness is reflected by a naʻau pono. Those with naʻau pēpē are praised for their modesty, while those with naʻau hoʻokiʻekiʻe, on the other hand, are ridiculed for their conceited nature. People who think too much, to the point that they have an indecisive nature, are known as naʻau lua, while those who are thoughtless are known for their naʻau kūhili. Even one's emotional stability emanates from the naʻau. In times of happiness, it is content, but when one has a naʻauʻauā due to intense grief, even suicide is possible. For Kānaka, the naʻau does not merely perform a bodily function; the naʻau defines people, right down to the heart and soul of their being.

Listening to our naʻau is just as important as listening to the sounds and seeing the sights around us. The naʻau helps the mind and body to make wise choices, and unlike the five previous sense abilities, which come naturally to most people, naʻau is not a sense ability that all people recognize. It is sharpened to a higher degree of acuity the more it is utilized and atrophies when it is not.

The Sense Ability of Kulāiwi

The sense ability of kulāiwi can be described as an intimate connection to place. Kula is a "plain or source," while iwi means "bone." New generations are born on our kulāiwi and the bones of our kūpuna are laid to rest there as well; thus, many Kānaka treasure the ʻāina for its ancestral ties. In fact, the iwi of our kūpuna are considered to be so sacred that they are the most cherished items in our care.[13] Tied to this concept is the word "ʻōiwi" (native). Similarly, Māori, the indigenous peoples of Aotearoa (New Zealand), use the word "iwi" to refer to their ancestral tribes. Iwi are identified by both their ancestral lineages and places. Members of a single iwi descend from common ancestors and often reside on the same ancestral homeland that their ancestors once did hundreds of years ago. From this indigenous perspective, only those who are genealogically related to the ʻāina are ʻōiwi.

The sense ability of kulāiwi is a very deep and profound connection to place and ancestry. Like the sense ability of naʻau, the sense ability of kulāiwi is both an ʻike kūhohonu and an ʻike kumu for Kānaka. To be ignorant about our kulāiwi is to have forgotten our ancestry and origins—the foundations of Kanaka identity. Our sense ability of kulāiwi is not merely our relationship to place, but it also links us to our ancestors of the past, the elders of the present, and the unborn generations of the future. Being born and raised on our kulāiwi root Kānaka to the places we call home, creating an inalienable sense of belonging that spans many generations. Thus, Kānaka are also Kānaka ʻŌiwi (native people of the land). Ancestral lands are our birthrights and legacies. We connect with our kūpuna who lived, worked, and played on the same ʻāina for many generations before us. We identify with our kulāiwi and kūpuna from which we descend; we are intimately bound to the birthplaces and burial grounds of our ancestors. We are connected to "the most vital physical, psychological, social and spiritual values of one's existence."[14] We have well-established, firmly planted roots.

Kānaka often know our kulāiwi very intimately. A kupa (native) of a kulāiwi might recite the names of the winds and rains of that place, tell you the average level of the stream flow for any given season, and direct you to places to gather various resources. Although another person may have been born and raised nearby in a different kulāiwi, their knowledge of the place would not be as intimate as that of a kupa of that particular kulāiwi. The farther away we go from our kulāiwi, the less we tend to know about places.

In modern society, however, there is often the misconception that Kanaka academics and cultural practitioners should know every minute detail about anything and everything Kanaka, including our history, culture, and geography. In actuality, no one knows the distinctive characteristics and geographies of *every* place. No single person living now knows *everything* there is to know about fishing, bird catching, weaving lau hala (pandanus leaves), pounding poi, beating kapa, and other ancestral practices either. In fact, the same can be said of our kūpuna. Ancestrally, Kānaka had very specialized skills; often, at birth or at a very young age, children were selected for a particular profession and were reared to be masters of their trades. Kanaka knowledge was often highly localized as well. Lawaiʻa (fishers) knew about the conditions of the ocean and methods of fishing that best suited the needs of their own places. While general knowledge of particular practices can be applied to other places in ka pae moku, some techniques are adapted generation after generation and are uniquely suited for specific needs within particular kulāiwi.

This sense ability of kulāiwi is an affinity with the ʻāina, especially the specific places we call home. Our kulāiwi inform our identities, so much so that knowing where a person is from is perhaps more important than even knowing that person's name. In fact, it is often considered rude to outwardly ask someone's name. A more appropriate question may be "No hea mai ʻoe?" (Where are you from?). References to our kulāiwi are embedded with information about a person and their upbringing. In essence, knowing where someone is from partially answers the question, "Who are you?" After stating the place from which you descend, a kupuna might ask, "Who's your family?" The family connection takes precedence over the individual. Identifying who our ancestors are and where our kulāiwi are located speaks volumes about who *we* are.

The sense ability of kulāiwi recognizes our deep kuleana (responsibility and burden) to care for the ʻāina that sustains us. When land tenure practices in Hawaiʻi shifted from communal stewardship to the privatization of land, the term "kuleana" was adopted for lands granted to the makaʻāinana. At a time when exact land boundaries were being surveyed and demarcated for privatization and foreigners were seeking to obtain lands to exploit for their personal gain, many Kānaka remained rooted to the same places their own kūpuna once resided upon. Since many Kānaka viewed the ʻāina as family rather than a commodity to be bought and sold, the term "kuleana" was applied to parcels of land granted to makaʻāinana. The makaʻāinana had the burden and responsibility to care for these kuleana lands in perpetuity.

The sense ability of kulāiwi celebrates our inalienable bonds with the ʻāina that feeds us and the kūpuna who have tilled the soil before us. Irrigation ditches, rock walls, heiau, and house lot foundations are footprints linking us to our kūpuna and our kulāiwi. As we accept our kuleana to serve as custodians of these family heirlooms, we ensure that the next generation will know their roots and develop a keen sense ability of kulāiwi. Current generations are links to the past, present, and future, so we have a kuleana to succeeding generations to absorb the ʻike of our kūpuna and to ensure that we train the next generation to be repositories of *our* ʻike kupuna.[15] The sense ability of kulāiwi is a Kanaka geography that grounds us and informs our identities by connecting us to our ancestral homelands and to the ancestors from whom we descend.

The Sense Ability of Au ʻĀpaʻapaʻa

The sense ability of au ʻāpaʻapaʻa (ancestral time) reflects ancestral ways of knowing and timekeeping. Kanaka knowledge is largely based on au ʻāpaʻapaʻa, extending back to the creation of the world.[16] As previously discussed, mele koʻihonua and other moʻolelo record this knowledge base and connect us genealogically to all life forms in our environment, even including the coral polyp, the first living organism in Kanaka traditions. The sense ability of au ʻāpaʻapaʻa is a native intelligence and science based on in-depth observation of the relationships between various species and the study of timeless constants in our environment.

Our system of timekeeping is based on naturally occurring phenomena, such as the rising and setting of the sun. We utilize the same sun, moon, and stars as our kūpuna once did to record natural time; these are our "grandfather clocks." Daily periods of time are divided into ao or ʻeleao (day) and pō or ʻelepō (night). Ao is further divided into kakahiaka (morning),[17] awakea (time when the sun is high in the sky), ʻauinalā (afternoon when the sun begins to decline), and ahiahi (evening).[18] Pō (night) is divided into pō and aumoe (late night—midnight hours).[19] The sun dawns at wanaʻao. The time when the sun is directly overhead and the shadow of a person retreats back into the body is referred to as "kau ka lā i ka lolo, hoʻi ke aka i ke kino."

Just as ancestral Kānaka observed solar cycles, they studied the lunar cycles too, noting correlations between the nights of the moon and the farming and fishing harvests associated with each of the nights. Over time, they developed sophisticated moon calendars to aid in their natural resource management practices for their particular kulāiwi. Moon calendars were oral in nature, with each phase of the moon receiving its own distinctive name based on its characteristics and/or its effect on the environment. The name "Hua" (egg, fruit, seed) was given to a night that resembles the shape of an egg and is noted for producing excellent fruit crops with bountiful harvests. The name "Huna" (hidden) was given for an excellent night to fish. On this particular night, marine life can be easily caught, as fish are prone to remaining in their holes. Kāloa Kū Kahi (Kāloa standing first) and Kāloa Kū Lua (Kāloa standing second) were productive nights for planting crops that farmers desired to grow long, such as kō (sugarcane), mai'a (bananas), 'ohe (bamboo), wauke (paper mulberry), and vines because embedded in the name Kāloa is the term "loa," meaning to be long. Unproductive farming and fishing nights were named "'Ole" (none, zero).

The moon also tracked the months and seasons of the year, with the names of the months and seasons varying from island to island.[20] According to Kamakau, the two seasons were Makali'i and Ho'oilo.[21] Others contend that the summer months were known as the season Kau, while the winter months were referred to as Ho'oilo.[22] During the summer months, farming, fishing, kapa beating, hunting, and gathering fruits were common practices.[23] Weaving baskets and mats, spinning cordage, and repairing tools and weapons were tasks typically completed indoors during the wetter winter months.[24]

The moon calendar was just one way in which ancestral Kānaka tracked time. Through careful observation of the natural environment over many generations, they also recognized that certain aspects of nature follow predictable and reliable seasons. Thus, the life cycles of plants and the mating cycles of animals also served as natural timekeepers. "Pua ka neneleau, momona ka wana," for example, is the time when "the neneleau blooms and the sea urchin is fat" (ON 295). For people living inland who cannot see marine life from their residences, the blooming of the neneleau plant alerts them that sea urchins are ready to be harvested. When "pua ke kō, kū ka he'e" (the the sugarcane tassels bloom, the octopus appears), this is an indication that it is late October or early November (ON 295). Welehu is the month when new leaf buds are forming on the 'ōhi'a trees, as alluded to in the saying "Welehu ka malama, liko ka 'ōhi'a" (ON 321). The month of Nana is the time in which fledglings leave their nests, and children born during this month are characterized as having an adventurous spirit (ON 306).[25]

Just as Kānaka recognized that many aspects of life follow a fairly predictable season, they also noted that people developed along a somewhat predictable timeline. Ancestrally, Kānaka did not place great emphasis on one's age in terms of years; rather, time constructs were based on one's ability level during the different phases or seasons of life. "Ka 'ōpu'u pua i mōhala" metaphorically makes reference to a baby by comparing it to "a flower that began to unfold" (ON 164). A child who is too young to know any better is said to be a "kamali'i 'ōkole helele'i" (a loose-seated child), who excretes anywhere indiscriminately (ON 159).

As children mature and are capable of completing new tasks, these milestones become indicators of their age. "Ka nui e pa'a ai i ka hue wai," refers to children who are able to carry a water gourd, about two years old. Around the age of five, when children are big enough to carry two coconuts, the phase of life is said to be "ka nui e pa'a ai i nā niu 'elua." Around the age of ten, children mature to "ka nui e 'auamo ai i ke keiki i ke kua" (the size that enables one to carry a smaller child on the back) (ON 163). A person who is physically or mentally childlike is said to have bones that are not yet matured in the saying, "Aia i ka mole kamali'i, 'a'ohe i o'o ka iwi" (ON 7). Later in life, when a boy is almost old enough to have sexual relations, he is said to be "kokoke e 'ā ke ahi o ka 'aulima" (almost ready to make fire with a fire stick held in the hand) (ON 197). When he is able to start a fire and can broil food, he reaches the phase in his life when he is old enough to have a mate: "ka nui e mo'a ai ka pūlehu" (ON 163). But when a person ages, the final season of one's life is compared to the setting of the sun in the phrase, "'O ka 'aui aku nō koe o ka lā" (ON 262). One's entire life span is referred to poetically as "mai ka hikina a ka lā i Kumukahi a ka welona a ka lā i Lehua" (from the rising of the sun in the east at Kumukahi until the setting of the sun in the west at Lehua) (ON 223–224).

The sense ability of au 'āpa'apa'a recognizes that Kanaka knowledge systems are based on grand ancestral time. Our 'ike is the result of thousands of years of scientific theory, observation, and practice. Our native intelligence has withstood the test of time and the scrutiny of each succeeding generation. The kūpuna of yesteryear and the youth of today all rely on the same points of reference to tell time, to navigate throughout the Pacific, and to determine the best nights to fish and farm. Thus far, our sense of au 'āpa'apa'a has proven to be timeless, remaining relevant and applicable to each generation.

The Sense Ability of Mo'o

Kanaka knowledge is a continuum passed down generation after generation, from master to apprentice, until the novice ultimately becomes a kahuna in his or her own right. With each succeeding generation in ka pae moku, Kānaka have refined and advanced the body of knowledge that our ancestors previously developed. This compilation of knowledge, passed down genealogically over many generations, is 'ike kupuna.

The Kohala, Waimea, and Kona field systems on the island of Hawaiʻi exemplify the extraordinary innovation and ancestral knowledge that our kūpuna developed over time. Remnants of rock walls are still visible on the landscape today, attesting to the intensive cultivation that our kūpuna practiced. Field systems were uniquely tailored to specific places. In dry and windy climates like Kohala, our kūpuna cultivated sugarcane and other crops to serve as permeable windbreakers, capturing fog, mist, and rainfall for surrounding crops. Modern scientific tests indicate that even in quite dry areas, ancestral planting methods created niches of damp soil necessary for crops to thrive. By planting complementary crops near one another, our kūpuna were able to grow staple crops in seemingly arid and undesirable soils. The level of cultivation was so extensive in the 1700s in the Kohala, Waimea, and Kona regions that Western explorers reported that our kūpuna cultivated their land to the fullest extent possible.

Highly specialized divisions of labor fostered innovation in ancestral times. Our kūpuna often focused their attention on a specific activity in a particular place. The construction of loʻi and loko iʻa, for example, reveals how ʻike became increasingly sophisticated over time. Throughout ka pae moku, Kānaka created irrigation systems that diverted freshwater from streams and fed wetland loʻi, although ensuring that the water flowed through each loʻi from the mountains to the sea was no easy engineering feat. Each community had topographies and water sources that were unique to their places. Careful calculations guaranteed that each loʻi was constructed at the right elevation to ensure adequate water flow for all loʻi, and once the loʻi was irrigated, the nutrient-rich water was then redirected back to the stream. In many instances, the mouth of the river emptied into loko iʻa specifically tailored to the river mouth and coastline of the place. The nutrient-rich water attracted fish to spawn in the loko iʻa. Once inside the mākāhā (sluice gates) of the loko iʻa, fish fed on the rich and abundant food supply. As fish matured and grew larger, they could no longer fit through the small mesh of the mākāhā and were trapped inside the loko iʻa. The level of skill and ingenuity needed to construct such irrigated loʻi and loko iʻa conceivably took many generations to perfect.

Central to Kanaka knowledge systems is the term "moʻo," defined by Pukui and Elbert as "succession, series, especially a genealogical line, lineage" and "grandchild."[26] This word is quite indicative of a Kanaka worldview that privileges the intelligence of the collective body, rather than that of a single brilliant individual. "Mo'o" is the prefix of many words, including but not limited to mo'okū'auhau (genealogy), mo'o ali'i (genealogy of ali'i), mo'o kanaka (succession of people), mo'o hele (path), mo'o ka'ao (story), and mo'olelo (historical account, literature, narrative).

Mo'o is a series, a culmination of insight gained and traditions practiced by the Kanaka community. We follow in the many footsteps of our ancestors; we know because our ancestors knew. For Kānaka and other indigenous peoples, culture, the mo'o of footsteps taken by our kūpuna, is a cornerstone of our societies. Each succeeding generation builds upon our culture, creating a solid foundation for the next. Kānaka of each succeeding generation learn the wisdom of our kūpuna, forming a knowledge base carefully maintained and expanded upon generation after generation. Through our mo'o, social practices are internalized and normalized.

The Kanaka knowledge continuum is also predicated on mo'okū'auhau.[27] Mo'o is a tradition handed down until ultimately reaching the mo'opuna, the grandchild and recipient of the pūnāwai waiwai (rich repository of wisdom) of our kūpuna. Central to both mo'opuna and kupuna is the word "puna" (metaphoric source of knowledge). By the time people earn the title of "kūpuna," they have had many life experiences, all of which contribute to their understanding of the world. The word "kūpuna" evokes the words "kū" (to stand) and puna (spring). Through their carefully selected haumāna, their legacy flows from themselves to the next generation as mo'o (continuous) puna (springs). Kūpuna are "sources of knowledge," and from them, the succeeding generations are able to kupu (to sprout forth and grow up).

Being able to mo'o helu (enumerate and recite) the legacies of their kūpuna brought a sense of pride to ancestral Kānaka, in turn identifying where they came from and who they were, carving out a niche for themselves in society. Our mo'o kupuna (ancestral genealogy) often predetermines what we become in the future, thus perpetuating the line of succession of our kūpuna. A person from a long line of kālai wa'a (canoe builders) would likely follow in the footsteps of his skilled kūpuna, carrying on the family tradition. Kānaka often remark, "He mēheuheu mai nā kūpuna" (Habits are acquired from one's ancestors) (ON 89). Mo'opuna serve as a link between the past and the future, ensuring that the bones of our ancestors and our mo'okū'auhau survive eternally.

Mo'okū'auhau were important in ancestral times because they revealed the accomplishments, extraordinary feats, and historical facts of one's kūpuna.

Moreover, they ordered Kanaka space and time by identifying the contemporaries and familial links of individuals.[28] Ancestrally, aliʻi and kāhuna memorized moʻo aliʻi, as one's lineage was a source of mana for descendants of high-ranking aliʻi. Offspring were in turn a source of mana for their ancestors; therefore, aliʻi strove to have moʻo lau (many descendants). By becoming powerful aliʻi in their own right, descendants increased the mana of their entire family line. Through the recitation of historical accounts honoring one's ancestors, the legacy of one's forefathers continued, thereby elevating the status of each succeeding generation.

Another way in which Kānaka honor our ancestors is reflected in the phrase "ola ka inoa" (the name lives) (ON 272). Ancestral names carried by pulapula (offspring, descendants) also provide mana to their namesake. Moreover, the process of naming a child after a kupuna allows the ancestor to be reborn through the descendant and to carry on the moʻolelo of that kupuna.[29]

Moʻolelo have the power to record history for succeeding generations. Before ʻōlelo Hawaiʻi became a written language, kāhuna memorized the moʻokūʻauhau of reigning chiefs as a means of validating a chief's right to rule, thereby ingraining this sentiment in the minds of their constituents and potential adversaries. Kāhuna had the responsibility of reciting moʻo aliʻi during important ceremonies. Since moʻokūʻauhau and other moʻolelo were a reflection of the aliʻi and their right to rule, failure to properly recite the names of important ancestors was potentially punishable by death. Makaʻāinana recited moʻolelo to their families and other makaʻāinana. Through moʻolelo, Kānaka were able to maintain a link to the past, describing the outstanding feats of one's ancestors, chronicling the events that happened at a particular locale, explaining the meanings of place names, and attributing the formation of certain land features to gods and the like.

Ancestral moʻolelo continue to be a source of knowledge for contemporary Kānaka. Today, we are able to gain insight about yesteryears not only by listening to our kūpuna recite the moʻolelo that have been passed down in the family but also by referring to written sources. Scholars are now able to access public and private collections that house audio and video recordings, as well as written sources in the form of books, journals, diaries, manuscripts, and maps. The oral and written documentation of Kanaka history affords scholars the opportunity to gain a wealth of knowledge about the past and to ensure that the legacy of our kūpuna is remembered by future generations.

As with ancestral Kānaka, for contemporary generations to know our future we must know our past. "Mua" and "hope" both mean the future and the past in ʻōlelo Hawaiʻi, illustrating a Kanaka perspective in which the past and the future are overlapping concepts. As we look back, we reflect on the teachings of our kūpuna and the ancestral knowledge they have imparted on succeeding generations. As the recipients of this legacy of ʻike, we move forward with the knowledge that the multitude of kūpuna that preceded us laid a firm foundation for us in the way of moʻokūʻauhau and moʻolelo.

Sensual Knowledge Systems

Kanaka knowledge systems are sensual in that we acquire ʻike by utilizing our sense abilities as repositories of knowledge. Throughout the day, we constantly rely on our multiple sense abilities, often subconsciously, to gain clarity about the world and the activities that we are engaged in. "Ma ka hana ka ʻike" is an ancestral proverb that suggests that learning is achieved through active participation and experience. Yet engaging in an activity in and of itself does not ensure knowledge acquisition. Enlightenment and mastery occurs only when people are able to harness the power of their sense abilities and apply them to a given art form. Master artists skillfully utilize all of their sense abilities in their trade. A kahuna lāʻau lapaʻau (master healer) hones her sense abilities of kulāiwi and au ʻāpaʻapaʻa to direct her to the best place and time of day to gather her medicinal plants. She relies on her ʻike kupuna and sense ability of moʻo to recall the teachings of her kumu (teacher, source of knowledge) and her sense abilities of sight, touch, taste, smell, hearing, and naʻau to ensure that her medicinal plants are the correct size, shape, age, and potency for her purposes.

Kanaka knowledge is largely performative in nature, and our bodies are conduits of knowledge.[30] We acquire knowledge by observing masters and by engaging in practices firsthand. The more we engage in an activity, the more attuned we become to the sense abilities associated with that given practice. Through extensive experience, the body internalizes and memorizes its role in these activities in order to perform such actions again at a later time. Reading books, watching YouTube how-to videos, and listening to kūpuna discuss their techniques may be helpful, but in order for someone to truly understand an art form it is necessary to have firsthand experience.

As modern Kānaka seek to relearn many of the traditions and customary practices of our kūpuna, we sometimes need to look no further than our own sense abilities. Our sense abilities of sight, listening, taste, touch, smell, naʻau, kulāiwi, au ʻāpaʻapaʻa, and moʻo allow us to tap into our ancestral memory. When we are silent and reconnect with our naʻau, ancestral memory, and environment, many of the answers that we are seeking suddenly manifest themselves. Hōʻailona that we failed to notice in the past unexpectedly appear. For the first time, we recognize the clues embedded in our ʻōlelo makuahine. From names of the moon phases to the ʻōlelo noʻeau describing ancient practices and observations, we are capable of sensing the knowledge we seek. It is up to us to sense the hōʻailona that our kūpuna have bequeathed to us so that we too may add new ʻike to our knowledge base for subsequent generations of Kānaka.

Conclusion

This book has taken us on a meandering path through ancestral time and space. It has been a journey of reconnection and enlightenment. Each bend in the road has presented a new chapter in the moʻolelo of a Kanaka sense ability of place and a Kanaka geography. Throughout this journey, our kūpuna have been escorting us on this path, linking us to our past, grounding us in the present, and paving a legacy for the future.

The ancestral knowledge systems of our kūpuna have guided us for countless generations. Ancestral indigenous ways of knowing are valid systems of knowledge grounded in the places from which they evolve. We must quote the moʻolelo of our ancestors and walk on this path so other indigenous peoples may likewise be inspired to embark on their own personal meandering journeys of rediscovery and enlightenment.

This alanui kīkeʻekeʻe (zigzag road) has reminded us that the path to Kanaka knowledge is not always straight and linear. But, like the alaloa o Maui (ancestral path around the island of Maui) that encircles the island, this path has no beginning or end: ancestral knowledge is a holistic continuum grounded in the past, relevant in the present, and indispensable in the future.[31]

As Kānaka, we have an obligation to our kūpuna to retain as much ʻike about ancestral ways of knowing as possible. We have a responsibility to be lovers of wisdom who are ready to accept the hikianakopili (spittle passed from a dying master) as it comes our way and to be the stewards of ancestral knowledges and ancestral places for future generations.

We have a kuleana to care for the ʻāina and the legacies created by our kūpuna. Highly sophisticated fishponds, agricultural irrigation systems, and grand heiau are all footprints that map Kanaka existence. When we care for these sites, we honor the heritage and ancestral places bequeathed to us over thousands of years by our kūpuna. While stewardship of this ʻike is an awesome kuleana, we are once again comforted by the words of aliʻi nui Liholiho: "Na wai hoʻi ka ʻole o ke akamai, he alanui i maʻa i ka hele ʻia e oʻu mau mākua?" (Who does not have the intelligence to travel a path that has been well-traversed by their ancestors?) (ON 251).

Notes

Introduction

1. My grandfather called this place Kuewa. According to the *Buke Mahele*, the place name is also spelled Kuewaa (perhaps Kuewaʻa using modern orthography).
2. At birth and death, the bond between the Kanaka and the ʻāina is physically and symbolically acknowledged. The relationship between the ʻāina and the newborn child, for example, is established through ancestry and the placing of the ʻiewe and piko of the child in the ground or in rocks of the Kanaka's kulāiwi. Metaphorically, the pōhaku (stone) in which the piko is placed provides a firm foundation for the child's life. The pōhaku is analogous to remaining steadfast in one place for a long period of time and thus becoming well acquainted with a place. The kulāiwi on which the pōhaku is situated represents the love and wisdom of the kūpuna who have previously walked the ʻāina. In this way, the baby and the ʻāina embrace one another and are connected by the life-giving piko. This relationship is further reinforced when a person's physical body dies and is kanu ʻia (planted); at death, burying a deceased person in the ʻāina brings this relationship full circle.
3. Throughout this book, I use "Kanaka" to refer to a Native Hawaiian person and "Kānaka" to refer to Native Hawaiians. Native Hawaiians include only descendants of the aboriginal people residing in ka pae ʻāina Hawaiʻi (the Hawaiian archipelago) prior to 1778. When not capitalized, "kanaka" and "kānaka" mean "person or human being" and "people or human beings" respectively. "Kanaka" is used as an adjective (e.g., Kanaka geographies).
4. Noenoe K. Silva, *Aloha Betrayed* (Durham and London: Duke University Press, 2004), 13.

5. Although Mary Kawena Pukui authored a number of ʻōlelo Hawaiʻi texts (e.g., *Hawaiian Dictionary, ʻŌlelo Noʻeau)* that ultilize ʻokina and kahakō, I have heard her family state that she did not want her name written with ʻokina and kahakō. In keeping with her wishes, her name appears throughout the book without ʻokina and kahakō.
6. For the purposes of this book, I rarely incorporate kaona into my writing. However, if I am successful at incorporating kaona when and where appropriate in the book, the reader may not even recognize this poetical use of the language.

CHAPTER 1: Mele Koʻihonua

1. J. M. Poepoe, "Ka Moolelo Hawaii Kahiko: Mokuna I: Na Kuauhau Kahiko e Hoike ana i na Kumu i Loaa ai ka Pae Moku o Hawaii nei," *Ka Na'i Aupuni,* February 6, 1906. In 1912, John H. Wise reprinted much of J. M. Poepoe's "Ka Moolelo Hawaii Kahiko: Mokuna I: Na Kuauhau Kahiko e Hoike ana i na Kumu i Loaa ai ka Pae Moku o Hawaii nei" in *Ke Au Hou* in a series of articles entitled, "He Moolelo No ka Hookumuia Ana o na Paemoku O Hawaii Nei ame ka Hoolaukanaka ana i Hoikeia ma na Mele Hawaii Kahiko." On January 17, 1912, the nineteenth anniversary of the overthrow of the Hawaiian monarchy, John H. Wise wrote, "Ke hoopuka aku nei makou ma keia wahi o KE AU HOU i ka Moolelo Kahiko o Hawaii nei, e like me ia i hoomakaukau ia ai a kakauia ai e JOSEPH MOKUOHAI POEPOE, a ke lana nei ko makou manao e lilo ana keia mahele i mea e pulama ia ai e na Opio Hawaii. Ua pili keia moolelo i ko Hawaii nei 'Ancient History,' elike me ia i hoike ia ai ma na mele a me na kuauhau a ka poe kahiko." (We are publishing the Ancient History of Hawaiʻi in KE AU HOU, just as JOSEPH MOKUOHAI POEPOE prepared it and wrote it, and we are hopeful that this will become a piece that is beloved by Hawaiian Youth. This historical account relates to the "Ancient History," as it appears in songs and genealogies of the ancient people.) Since Wise's articles quote J. M. Poepoe, citations in this book will credit J. M. Poepoe, the original author of this text.
2. More information exists on the *Kumulipo* than any other mele koʻihonua; thus, a greater discussion of this mele koʻihonua appears here.
3. S. L. Peleioholani, "The Ancient History of Hookumu-Ka-Lani = Hookumu-Ka-Honua," in Bishop Museum Archives (HI.L.1.3.1), 3.
4. Rubellite Kawena Johnson, *Kumulipo: Hawaiian Hymn of Creation,* vol. 1 (Honolulu: Topgallant Publishing Co. Ltd., 1981), 27.
5. "Moolelo Hawaii," *Ka Hoku o Hawaii*, December 21, 1911.
6. Liliuokalani, *The Kumulipo: An Hawaiian Creation Myth* (Kentfield: Pueo Press, 1997), introduction.

7. Ibid., preface, introduction.
8. Queen Lili'uokalani was imprisoned for having knowledge of a counterrevolutionary attempt by her supporters against the illegal provisional government.
9. Liliuokalani, *Kumulipo,* introduction.
10. The *Kumulipo* was originally recorded orally.
11. Liliuokalani, *Kumulipo,* introduction.
12. Ibid.
13. Ibid., 1–22.
14. Malo states that Kealiiwahilani was the husband of La'ila'i. David Malo, *Hawaiian Antiquities: Mo'olelo Hawai'i* (Honolulu: Folk Press, 1987), 4; "Moolelo Hawaii," *Ka Hoku o Hawaii*, December 21, 1911.
15. Liliuokalani, *Kumulipo,* 22–66.
16. Ibid., 66.
17. Martha Warren Beckwith, *The Kumulipo: A Hawaiian Creation Chant* (Honolulu: University of Hawai'i Press, 1972), 231.
18. Abraham Fornander, *Fornander Collection of Hawaiian Antiquities and Folk-Lore: The Hawaiian Account of the Formation of Their Islands and Origin of Their Race, with the Traditions of Their Migrations, Etc., as Gathered from Original Sources,* ed. Thomas G. Thrum, Facsimile ed., vol. 4 (Honolulu: 'Ai Pōhaku Press, 1999), 20–21.
19. Kahikoluamea and Kupulanakēhau are previously introduced in the twelfth wā. Liliuokalani, *Kumulipo*, 66.
20. Liliuokalani, *Kumulipo*, 74–75. An interesting aspect of the *Kumulipo* is that Maui and Hawai'i chiefs generally trace their lineage back to 'Ulu rather than his brother Nānā'ulu.
21. Martha Warren Beckwith, *Hawaiian Mythology* (Honolulu: University of Hawai'i Press, 1970), 226.
22. Liliuokalani, *Kumulipo,* 70–78.
23. Bishop Museum Archives, "HI.L1.3#5."; Peleioholani, "Ancient History," 3.
24. Papa had many different names, including Papahānaumoku and Haumea. Bishop Museum Archives, "HI.L1.3#4." Kamakau says her names were Papanuihānaumoku, Haumeahānauwāwā, Kahakauakoko, Haiuli, Laumiha, Kāmeha'ikana. Samuel Mānaiakalani Kamakau, *Tales and Traditions of the People of Old: Nā Mo'olelo O Ka Po'e Kahiko* (Honolulu: Bishop Museum Press, 1993), 134. Her parents, Kukalaniehu and Kahakuakoko, were said to be cliffs. Bishop Museum Archives, "HI.L1.3#4."
25. Wākea was also known as Loloimehani. Bishop Musuem Archives, "HI.L1.3#4."
26. Liliuokalani, *Kumulipo,* 66, 70.
27. Lilikalā Kame'eleihiwa, *Native Land and Foreign Desires: How Shall We Live in Harmony: Ko Hawai'i 'Āina a Me Nā Koi Pu'umake a Ka Po'e Haole: Pehea Lā e Pono Ai* (Honolulu: Bishop Museum Press, 1992), 23.

28. According to some sources, there is no place in Hawai'i named Loloimehani. "Moolelo Hawaii," *Ka Hoku o Hawaii*, December 21, 1911; Kamakau, *Tales and Traditions*, 128–129; S. M., Kamakau, "Ka Moolelo Hawaii: Mokuna 1," *Ke Au Okoa*, October 14, 1869; S. M. Kamakau, "Ka Moolelo Hawaii: Helu 2," *Ke Au Okoa*, October 21, 1869. However, some of these very sources contradict themselves by stating that Loloimehani was an ancient name for the island of O'ahu.
29. Fornander, *Fornander Collection*, vol. 4, 12–13.
30. Malo contends that Haumea and Papa are one and the same (page 5). Rubellite Kawena Johnson, *Essays in Hawaiian Literature*, Part 1: *Origin Myths and Migration Traditions* (Honolulu: University of Hawai'i, 2001), 35.
31. Kamakau, *Tales and Traditions*, 129, 134; Liliuokalani, *Kumulipo*, 66. Ho'ohōkūkalani per Kame'eleihiwa, *Native Lands and Foreign Desires*, 23.
32. Malo, *Hawaiian Antiquities*, 244.
33. Kame'eleihiwa, *Native Land and Foreign Desires*, 24.
34. In genealogy charts, "(w)" stands for "wahine" or "female" and "(k)" for "kāne" or "male." Ibid., 47.
35. Bishop Museum Archives, "HI.L8."
36. The twelve islands of the pae moku o Hawai'i are Hawai'i, Maui, Kaho'olawe, Lāna'i, Molokini, Moloka'i, O'ahu, Kaua'i, Ni'ihau, Lehua, Ka'ula, and Nihoa. Kame'eleihiwa spells the island "Māui" because it was named after the demigod by the same name. Throughout this book, I spell the island "Maui," without a kahakō, because it is no longer pronounced with the stress over the first "a" by native speakers of the 'ōlelo makuahine of these islands. Poepoe, "Ka Moolelo Hawaii Kahiko," *Ka Na'i Aupuni*, February 2, 1906.
37. Fornander, *Fornander Collection*, vol. 4, 3.
38. Peleioholani, "Ancient History," 8–14; Poepoe, "Ka Moolelo Hawaii Kahiko," *Ka Na'i Aupuni*, February 7, 1906.
39. Fornander, *Fornander Collection*, vol. 4, 12–13.
40. According to Fornander, the name was Keapapanui. Fornander, *Fornander Collection*, vol. 4, 13, 17. Ke'āpapanui is likely the spelling using Hawaiian orthography.
41. Fornander, *Fornander Collection*, vol. 4, 12–13, 16–17.
42. Poepoe, "Ka Moolelo Hawaii Kahiko," *Ka Na'i Aupuni*, February 3, 1906; Fornander, *Fornander Collection*, vol. 4, 12–13.
43. Fornander, *Fornander Collection*, vol. 4, 12–13. Mololani is probably another name for the island of Molokini. Johnson, *Essays in Hawaiian Literature*, 32.
44. Fornander, *Fornander Collection*, vol. 4, 12–19.
45. According to page 6 of *Mauiloa*, Kahitikū means "the eastern horizon," while Kahitimoe means "the western horizon." According to Fornander, Tahiti-ku

means "Tahiti of the rising sun" and Tahiti-moe means Tahiti of the setting sun." Fornander, *Fornander Collection*, vol. 4, 12–13. David Malo grew up in the royal court of Kamehameha I, worshipping ancestral gods and living under the kapu system. Malo's perspective on life is key to understanding the transitional period between traditional times and those heavily influenced by foreigners. In *Hawaiian Antiquities*, Malo notes the following:

15. The circle or zone of the earth's surface, whether sea or land, which the eye traverses in looking to the horizon is called *kahikimoe.*

16. The circle of the sky which bends upwards from horizon is *kahiki-ku;* above *kahiki-ku* is a zone called *kahiki-ke-papa-nuu;* and above that is *kahiki-ke-papa-lani;* and directly over head is *kahiki-kapui-holani-ke-kuina*" (10).

46. According to page 6 of *Mauiloa*, Ke'āpapanu'u means "the lower dome," while Ke'āpapalani means "the celestrial dome." Fornander uses Keapapanui and Keapapalani in his rendition. Ke'āpapanui and Ke'āpapalani are likely the spellings using Hawaiian orthography. Fornander translates these terms as "the foundation stones" and "the heavenly stones" respectively. Fornander, *Fornander Collection*, vol. 4, 12–13, 17.
47. "[O]lolani" (acclaimed chief) per Poepoe, "Ka Moolelo Hawaii Kahiko," *Ka Na'i Aupuni,* February 3, 1906.
48. "He nuu no Ololani, no Lono, no Ku" per Poepoe, "Ka Moolelo Hawaii Kahiko," *Ka Na'i Aupuni,* February 3, 1906. This line varies significantly from Fornander's version. Whereas Poepoe's version refers to a summit or a high-ranking chief, Fornander's suggests that Mololani was of great importance for Kū, Lono, Kāne, and Kanaloa. Kawena Johnson believes Mololani may be another name for Molokini. Johnson, *Essays in Hawaiian Literature,* 304, 325.
49. Malo, *Hawaiian Antiquities*, 243.
50. "Moolelo Hawaii,"*Ka Hoku o Hawaii,* December 21, 1911*;* Peleioholani, "Ancient History," 5; Fornander, *Fornander Collection*, vol. 4, 12–13.
51. Kamakau, "Ka Moolelo Hawaii: Mokuna 1," *Ke Au Okoa*, October 14, 1869; Kamakau, *Tales and Traditions*, 125; Peleioholani, "Ancient History," 5.
52. Fornander, *Fornander Collection,* vol. 4, 19, 21.
53. Ibid., 18–21.
54. Beckwith, *Hawaiian Mythologies,* 41.
55. Poepoe, "Moolelo Kahiko," *Ka Na'i Aupuni,* February 17, 1906. It is interesting to note that according to Beckwith, Lono and Kū are not mentioned in the *Kumulipo*. Beckwith, *The Kumulipo,* 171.
56. Fornander, *Fornander Collection,* vol. 6, 335. In the *Honua'ula* genealogy, the head of the first man was made of clay that Lono brought from the four corners of the world—Kai Ko'olau, Kai Kona, Kahikikū, Kahikimoe (North, South,

East, West)—while his body was made from red dirt. It took the efforts of three gods—Kāne, Kū, and Lono—to create Honuaʻula, who lived in Paliuli, a place formerly known as Kalana i Hauola. The gods then created a wife by the name of Keolakūhonua for him out of his ribs. Fornander, *Fornander Collection*, vol. 6, 267.

57. Kamakau, *Tales and Traditions*, 131.
58. Beckwith, *Hawaiian Mythology*, 564. According to Samuel Kamakau in *Ka Poʻe Kahiko: The People of Old*, Hulihonua and Keakahulilani were the first man and woman, respectively, of Hawaiʻi. Therefore, they are the ancestors of the Kānaka. Kamakau, *Ka Poʻe Kahiko*, 3.
59. Fornander, *Fornander Collection*, vol. 6, 335.
60. Kamakau, *Tales and Traditions*, 131.
61. Ibid.
62. Varying accounts credit different individuals with being the first man; according to Barrère, these include Kānehulihonua, Kumuhonua, Kuluipo, Kumuuli, and Hulihana. See also note 65.
63. Samuel Mānaiakalani Kamakau was born in 1815 at Manuaʻula, Oʻahu. When he was seventeen years old, he attended Lahainaluna, a missionary high school. There he became a teacher's assistant and later wrote articles about Kanaka culture and history. Kamakau was able to interview kūpuna who revealed many secrets about life in Hawaiʻi. Beginning in 1842, his work appeared in ʻōlelo makuahine newspapers on topics such as the coming of foreigners, onset of foreign diseases that decimated many Kānaka, deeds of Kamehameha, intricacies of the inner workings of Kanaka society, and the like. Although Kamakau is considered one of the best Kanaka historians to ever live, he did not write for a living. In his spare time, Kamakau explored his passion, recording some of the most important written accounts of Kanaka culture and history. Because of the vast knowledge contained in his ʻōlelo makuahine writings, his work has been translated and published in English. Samuel Mānaiakalani Kamakau, *Ruling Chiefs of Hawaii*, rev. ed. (Honolulu: Kamehameha Schools Press, 1992), iii–v.
64. Kamakau, "Ka Moolelo Hawaii," *Ke Au Okoa*, October 14, 1869; Kamakau, *Tales and Traditions*, 125; Malo, *Hawaiian Antiquities*, 3.
65. Beckwith, *Hawaiian Mythology*, 314–315.
66. Fornander, *Fornander Collection*, vol. 6, 335.
67. Kepelino and Martha Warren Beckwith, *Kepelino's Traditions of Hawaii*, Bernice P. Bishop Museum Bulletin 95 (Honolulu: Krause Reprint, 1971), 14–17.
68. Fornander, *Fornander Collection*, vol. 6, 335.
69. Kaulia, "He Adamu No Iloko O Ka Lahui Hawaii," *Ke Aloha Aina*, June 6, 1896.
70. Kamakau, *Tales and Traditions*, 32–33. The full names of these three gods are Kānenuiākea, Kūnuiākea, and Lononuiākea. Collectively they were known as

Kapukahi. Kaulia, "He Adamu No Iloko O Ka Lahui Hawaii," *Ke Aloha Aina,* June 6, 1896.

71. A similar account is given by Kaulia in "He Adamu No Iloko O Ka Lahui Hawaii," *Ke Aloha Aina,* June 6, 1896.
72. Kamakau, *Tales and Traditions*, 32–35.
73. Her name is Oure in "He Adamu No Iloko O Ka Lahui Hawaii," *Ke Aloha Aina,* June 6, 1896.
74. Barrère, "Cosmogonic Genealogies of Hawaii," 419.
75. According to Malo, the name of this genealogy is *Lolo.* Malo, *Hawaiian Antiquities,* 4. Liliuokalani and Kamakau call the genealogy *Ololo.* Liliuokalani, *Kumulipo,* 65; Kamakau, *Tales and Traditions*, 128.
76. Liliuokalani, *Kumulipo*, 65.
77. Kamakau, "Ka Moolelo Hawaii," *Ke Au Okoa*, October 14, 1869; Kamakau, *Tales and Traditions,* 132.
78. Kamakau, *Tales and Traditions*, 35.
79. This is quoted exactly as it appears in the original text, *Fornander Collection,* vol. 4, 21; thus, the Hawaiian orthography may differ from modern convention.
80. Kamakau, "Ka Moolelo Hawaii: Mokuna 1," *Ke Au Okoa*, October 14, 1869; Kamakau, *Tales and Traditions,* 128–129; Bishop Museum Archives, "HI.L1.3#4."
81. According to an unpublished work in the Bishop Museum Archives, "HI.L1.3#4," the residence of Kahiko and his wife was Kamawailualani. In "Ka Moolelo Hawaii," *Ke Au Okoa*, October 14, 1869 and Malo's *Hawaiian Antiquities* (page 5), the place name was recorded as "Kamawaelualani." "Ka-māwae-lua-lani" appears in Kamakau's *Tales and Traditions of the People of Old: Nā Moʻolelo a ka Poʻe Kahiko*, 128–129. "Kamāwaelualani" is an ancient name for the island of Kauaʻi. Kamakau, "Ka Moolelo Hawaii: Mokuna 1," *Ke Au Okoa*, October 14, 1869; Kamakau, *Tales and Traditions*, 129.
82. Wākea was also known as Loloimehani. Bishop Museum Archives, "HI.L1.3#4."
83. Kamakau, *Tales and Traditions,* 35. According to James Kaulia in "He Adamu No Iloko O Ka Lahui Hawaii," *Ke Aloha Aina*, June 6, 1896, this child was known as Makalu.
84. Kamakau, "Ka Moolelo o Hawaii Nei," *Ka Nupepa Kuokoa,* July 29, 1865. See also Kamakau, *Tales and Traditions,* 35.
85. Fornander, *Fornander Collection,* vol. 4, 22–23. See pages 24–25 for a genealogical chart of the ʻŌpuʻukahonua line.
86. Beckwith, *Hawaiian Mythology,* 307.
87. Fornander, *Fornander Collection,* vol. 4, 24.
88. Ibid., 3.
89. Ibid., 4.

90. Ibid.
91. Ibid.
94. Malo, *Hawaiian Antiquities*, 3; Wise, "He Moolelo No Ka Hookumuia," *Ke Au Hou*, January 24, 1912.
93. Bishop Museum Archives, "Moolelo"; Wise, "He Moolelo No Ka Hookumuia," *Ke Au Hou*, January 24, 1912, "Mookuauhau Alii: Na Iwikuamoo o Hawaii Nei mai Kahiko Loa," *Ka Makaainana*, June 8, 1896.
94. Liliuokalani, *Kumulipo*, 23, 60.
95. Kamakau, "Ka Moolelo Hawaii: Mokuna 1," *Ke Au Okoa*, October 14, 1869; Kamakau, *Tales and Traditions*, 125.
96. Mary Kawena Pukui, *ʻŌlelo Noʻeau: Hawaiian Proverbs and Poetical Sayings* (Honolulu: Bishop Museum Press, 1983), 183. Hereafter cited in text as "ON."
97. Malo, *Hawaiian Antiquities*, 2.
98. Beckwith, *Kumulipo*, 153.
99. Kameʻeleihiwa, *Native Land and Foreign Desires*, 26.

CHAPTER 2: Places to Stand

1. George S. Kanahele, *Kū Kanaka, Stand Tall: A Search for Hawaiian Values* (Honolulu: University of Hawaiʻi Press, 1986), 194–202.
2. Mary Kawena Pukui, E. W. Haertig, and Catherine A. Lee, *Nānā I Ke Kumu (Look to the Source)*, vol. 1 (Honolulu: Hui Hanai, 1983), 91–92.
3. Ibid., 27–28.
4. Pukui, *ʻŌlelo Noʻeau*, 243. Hereafter cited in text as "ON."
5. "Ka Moolelo o Hawaii Nei: Helu 8," *Ka Nupepa Kuokoa*, August 5, 1865; Kamakau, *Tales and Traditions*, 35. Others have enumerated the social strata of ancient Hawaiʻi differently. According to Handy, Handy, and Pukui in *Native Planters*, the society was divided into four categories: papa aliʻi (chiefs), koa (warriors), poʻe akamai o ka ʻāina (people who were knowledgeable about the land), and poʻe pale ʻike (keepers of knowledge). E. S. Craighill Handy, Elizabeth Green Handy, and Mary Kawena Pukui, *Native Planters in Old Hawaii: Their Life, Lore, and Environment* (Honolulu: Bishop Museum Press, 1972), 46.
6. Kanahele, *Kū Kanaka, Stand Tall*, 196–197.
7. The word "akua" has multiple, meanings, including "god, goddess, spirit, ghost, devil, image, idol, corpse; divine, supernatural, godly." Pukui and Elbert, *Hawaiian Dictionary: Hawaiian-English, English-Hawaiian* (Honolulu: University of Hawaiʻi Press, 1986), 15. Per Pukui and Elbert, "Lesser chiefs who served higher chiefs were called kauā maoli" (page 134) or "aliʻi poʻe kauā" (page 20).
8. Kanahele, *Kū Kanaka, Stand Tall*, 196.

9. Carlos Andrade, *Hā'ena: Through the Eyes of the Ancestors* (Honolulu: University of Hawai'i Press, 2008), 75–76.
10. Beckwith, *Hawaiian Mythology*, 376.
11. Kanahele, *Kū Kanaka, Stand Tall,* 196.
12. Kauikeaouli and Nāhi'ena'ena were the last nī'aupi'o. J. Waiamau, "Ka Hoomana Kahiko Helu 27: Na Papa O Kanaka, Na'lii, Na Makaainana, Na Lopa, Na Hu, Na Kauwa," *Ka Nupepa Kuokoa,* November 11, 1865.
13. For a more in-depth discussion of ali'i nui, please refer to: S. M. Kamakau, "Ka Moolelo Hawaii: Helu 46," *Ke Au Okoa,* October 27, 1870; S. M. Kamakau, *Tales and Traditions of the People of Old;* Malo, *Hawaiian Antiquities;* Ka Moolelo o Hawaii Nei: Helu 8," *Ka Nupepa Kuokoa,* August 5, 1865; "Na Lei o Hawaii," *Ka Nupepa Kuokoa,* February 24, 1883; Pukui and Elbert, *Hawaiian Dictionary.*
14. Fornander, *Fornander Collection,* vol. 6, 308. John Papa Ii defines "nī'aupi'o" slightly differently, including the offspring of first cousins. His definition is "a chiefly rank; the offspring of a union of chiefs who were siblings or first cousins." John Papa Ii, *Fragments of Hawaiian History,* ed. Dorothy B. Barrère, trans. Mary K. Pukui, rev. ed. (Honolulu: Bishop Museum Press, 1983), 182.
15. Kamakau, *Ka Po'e Kahiko,* 4.
16. Beckwith, *Kumulipo,* 13.
17. "Na Lei o Hawaii," *Ka Nupepa Kuokoa,* February 24, 1883; "Na Papa Alii o Hawaii," *Ka Nupepa Kuokoa,* March 10, 1883; Kamakau, "Ka Moolelo Hawaii: Helu 46," *Ke Au Okoa,* October 27, 1870; Kanahele, *Kū Kanaka, Stand Tall,* 196.
18. Kamakau, *Ka Po'e Kahiko,* 4–5; Malo, *Hawaiian Antiquities,* 55; Pukui and Elbert, *Hawaiian Dictionary,* 258.
19. Beckwith, *Kumulipo,* 13. The kapu noho was a taboo that required all people in the presence of the ali'i naha (chiefs of naha rank) to crouch in honor of the chiefs' sacredness. Anyone who failed to honor such a taboo was sentenced to death.
20. Fornander, *Fornander Collection,* vol. 6, 308.
21. Kamakau, "Ka Moolelo Hawaii: *Ke Au Okoa,* October 27, 1870"; Kamakau, *Ka Po'e Kahiko,* 5.
22. Ibid.
23. "Ka Moolelo o Hawaii Nei: Helu 8," *Ka Nupepa Kuokoa,* August 5, 1865; Kamakau, *Tales and Traditions,* 39–40.
24. Fornander, *Fornander Collection,* vol. 6, 309.
25. Kame'eleihiwa, *Native Land and Foreign Desires,* 45–47.
26. Ibid., 20–22, 43.
27. Ibid.
28. "Ka Moolelo o Hawaii Nei: Helu 14 No Ke Kaapuni Makaikai I Na Wahi Kaulana A Me Na Kupua, A Me Na'lii Kahiko Mai Hawaii a Niihau," *Ka Nupepa Kuokoa,* September 30, 1865.

29. "Mookuauhau Alii: Na Iwikuamoo o Hawaii Nei Mai Kahiko Mai," *Ka Makaainana*, May 25, 1896.
30. I believe his name should be spelled "Kanikani'ā'ula," which can mean "insomnia" or "to mourn in chanting." Pukui and Elbert, *Hawaiian Dictionary,* 129.
31. "He Mau Mea i Hoohalahala ia no na Mea iloko o na Kaao Hawaii," *Ka Nupepa Kuokoa,* February 15, 1868.
32. I believe his name should be spelled "Ka'ihiwālua," meaning "the sacred one carried on a pole on the shoulders of two people" because of his high-ranking genealogy. Pukui and Elbert, *Hawaiian Dictionary*, 94, 381.
33. "Ka Moolelo o Hawaii Nei," *Ka Nupepa Kuokoa,* September 30, 1865; "Mookuauhau Alii," *Ka Makaainana*, May 25, 1896, and July 6, 1896.
34. "Mookuauhau Alii," *Ka Makaainana,* May 25, 1896 and July 6, 1896.
35. "Ka Moolelo o Hawaii Nei," *Ka Nupepa Kuokoa*, September 30, 1865; "Mookuauhau Alii," *Ka Makaainana,* May 25, 1896.
36. Kame'eleihiwa, *Native Land and Foreign Desires,* 43.
37. "Ka Moolelo o Hawaii Nei," *Ka Nupepa Kuokoa*, September 30, 1865; "Mookuauhau Alii," *Ka Makaainana*, May 25, 1896, and July 6, 1896.
38. Kamakau, *Ka Nupepa Kuokoa,* September 30, 1865, and *Ka Nupepa Kuokoa,* July 26, 1901.
39. "Ka Moolelo o Hawaii Nei," *Ka Nupepa Kuokoa*, September 30, 1865; "Mookuauhau Alii," *Ka Makaainana*, May 25, 1896, and July 6, 1896.
40. "Ka Moolelo o Hawaii Nei," *Ka Nupepa Kuokoa*, September 30, 1865; "Mookuauhau Alii," *Ka Makaainana*, July 6, 1896.
41. "Mookuauhau Alii," *Ka Makaainana*, May 18, 1896.
42. Ibid., June 8, 1896; Kamakau, "Ka Moolelo Hawaii: Helu 58," *Ke Au Okoa,* December 15, 1870; Kamakau, "Ka Moolelo Hawaii: Helu 51," *Ke Au Okoa,* December 22, 1870. The marriage of Pi'ikea of Maui to 'Umialiloa was a very wise political move because it brought together the ali'i nui lines of Maui and the 'Ī and Keawe lines of Hawai'i Island.
43. Kamakau, *Ruling Chiefs,* 31–32.
44. "Mookuauhau Alii," *Ka Makaainana*, July 6, 1896.
45. Ibid., June 8, 1896, and July 6, 1896.
46. Ibid.
47. Ibid., June 8, July 6, and July 13, 1896.
48. Ibid., May 25, 1896.
49. Keku'iapoiwa Nui I of Maui, mother of Kamehamehanui, Kalola, and Kahekili, should not be confused with Keku'iapoiwanui of Hawai'i Island, mother of Kamehameha I (also known as Kamehameha Nui). "Ka Moolelo o Kamehameha I," *Ka Nupepa Kuokoa*, October 20, 1866.

50. "Mookuauhau Alii," *Ka Makaainana*, November 16, 1896; "Ka Mookuauhau Alii a WM. U. Hoapili Kanehoa," *Ka Nupepa Kuokoa*, August 15, 1913; Edith Kawelohea McKinzie, *Hawaiian Genealogies: Extracted from Hawaiian Language Newspapers*, vol. 2 (Honolulu: University of Hawai'i Press, 2003), 37.
51. "Mookuauhau Alii," *Ka Makaainana*, April 27, 1896, June 22, 1896, and November 16, 1896; "Ka Moolelo o Kamehameha I," *Ka Nupepa Kuokoa*, October 20, 1866.
52. "Mookuauhau Alii," *Ka Makaainana*, November 16, 1896; "Ka Moolelo o Kamehameha I," *Ka Nupepa Kuokoa*, October 20, 1866.
53. Kamakau, *Ruling Chiefs*, 188–189; "Ka Moolelo o Kamehameha I," *Ka Nupepa Kuokoa*, July 20, 1867.
54. "Ka Moolelo o Kamehameha I: Helu 11," *Ka Nupepa Kuokoa*, January 19, 1867; Kame'eleihiwa, *Native Land and Foreign Desires*, 72.
55. Fornander, Abraham, *An Account of the Polynesian Race: Its Origins and Migrations, and the Ancient History of the Hawaiian People to the Times of Kamehameha I* (Whitefish, MT: Kessinger Publishing, 2007), 209–210; Hawaiian Historical Society, *Annual Report of the Hawaiian Historical Society* (Honolulu: Hawaiian Historical Society, 1904), 14; Kamakau, *Ruling Chiefs*, 128; "Ka Moolelo o Kamehameha I," *Ka Nupepa Kuokoa*, March 16, 1867.
56. Kamakau, *Ruling Chiefs*, 128, 137; "Ka Moolelo o Kamehameha I," *Ka Nupepa Kuokoa*, March 16, 1867, and March 30, 1867; Kame'eleihiwa, *Native Land and Foreign Desires*, 47.
57. Kame'eleihiwa, *Native Land and Foreign Desires*, 46.
58. Kamakau, *Ruling Chiefs*, 69; "Mookuauhau Alii," *Ka Makaainana*, May 25, 1896; "Ka Moolelo o Kamehameha I," *Ka Nupepa Kuokoa*, October 20, 1866.
59. Kame'eleihiwa, *Native Land and Foreign Desires*, 20–22.
60. Liliuokalani, *Kumulipo*, preface.
61. Kamakau, *Ruling Chiefs*, 69; "Mookuauhau Alii," *Ka Makaainana*, November 16, 1896; "Ka Moolelo o Kamehameha I," *Ka Nupepa Kuokoa*, October 20, 1866.
62. Maka'āinana were free to move if they were not well cared for by their konohiki. Andrade, *Hā'ena*, 75–76.
63. Kanahele, *Kū Kanaka, Stand Tall*, 181.
64. Pukui and Elbert, *Hawaiian Dictionary*, 242.
65. Samuel M. Kamakau, *Works of the People of Old: Nā Hana a Ka Po'e Kahiko*, ed. Dorothy B. Barrère, trans. Mary K. Pukui (Honolulu: Bishop Museum Press, 1976), 5. For a more in-depth discussion of kūkulu, please refer to Kamakau, "Ka Moolelo Hawaii: Helu 4," *Ke Au Okoa*, November 4, 1869.
66. Bishop Museum Archives (MS SC Emory Group 8, Box 4.7); Bishop Museum Archives (HEN vol. 1, Hms K9 Kalanianaole Collection), 277–278.
67. Bishop Museum Archives (MS SC Handy, Esc, Box 7.30, Poepoe Collection).

68. Pukui, Haertig, and Lee, *Nānā I Ke Kumu,* vol. 1, 54, and vol. 2, 270.
69. Andrade, *Hā'ena*, 70.
70. Pukui and Elbert, *Hawaiian Dictionary,* 9, 11, 167–168.
71. Ibid., 9.
72. Ibid., 167–168.
73. Handy, Handy, and Pukui, *Native Planters in Old Hawaii,* 289.
74. *Ka Nupepa Kuokoa*, August 22, 1868.
75. Mary Kawena Pukui, Samuel H. Elbert, and Esther T. Mookini, *Place Names of Hawaii,* rev. and enl. ed. (Honolulu: University Hawai'i Press, 1976), 48.
76. "No ka Wehewehe Ana," *Ka Hae Hawaii*, March 12, 1856; "He Kanikau no ko Maua Papa Heleloa," *Ka Nupepa Kuokoa*, August 21, 1924; 'Ī'ī, *Fragments of Hawaiian History*, 159. Similarly, "Mauiokama" refers to Maui and "Mauinuiokama" refers to Maui along with the neighboring islands of Molokini, Kaho'olawe, Moloka'i, and Lāna'i.
77. Kamakau, *Ruling Chiefs,* 129; Kamakau, "Ka Moolelo o Kamehameha I," *Ka Nupepa Kuokoa*, March 16, 1867.
78. Kamakau, *Ruling Chiefs,* 137; "Ka Moolelo o Kamehameha I," *Ka Nupepa Kuokoa*, March 30, 1867.
79. Kamakau, "Ka Moolelo Hawaii: Helu 51," *Ke Au Okoa*, December 1, 1870.
80. Kamakau, "Ka Moolelo Hawaii: Helu 52," *Ke Au Okoa*, December 8, 1870.
81. Handy, Handy, and Pukui, *Native Planters in Old Hawaii,* 277.
82. Andrade, *Hā'ena*, 90; Kame'eleihiwa, *Native Land and Foreign Desires,* 51–52, 110.
83. Kame'eleihiwa, *Native Land and Foreign Desires,* 51–52.
84. Kaua'i was divided into five moku. Hawai'i and O'ahu were each divided into six moku. Pukui, Elbert, and Mookini, *Place Names of Hawaii,* xviii–xxi.
85. Kame'eleihiwa, *Native Land and Foreign Desires,* 29.
86. Andrade, *Hā'ena*, 74–75, 89; Paul F. Nahoa Lucas, *A Dictionary of Hawaiian Legal Land-Terms* (Honolulu: Native Hawaiian Legal Corporation, University of Hawai'i Committee for the Preservation and Study of Hawaiian Language, Art, and Culture, 1995), 57.
87. Andrade, *Hā'ena*, 75; Kame'eleihiwa, *Native Land and Foreign Desires,* 295.
88. Handy, Handy, and Pukui, *Native Planters in Old Hawaii,* 58.
89. Kanahele, *Kū Kanaka, Stand Tall,* 196.
90. J. Davies, A *Tahitian and English Dictionary* (Tahiti: London Missionary Society's Press, 1991), 136; personal communication, Lilikalā Kame'eleihiwa, March 2, 2006.
91. Pukui and Elbert, *Hawaiian Dictionary,* 168.
92. Ibid., 124, 184.
93. Ibid., 43.
94. Lucas, *Dictionary of Hawaiian Legal Land-Terms*, 12, 84.

95. Kanahele, *Kū Kanaka, Stand Tall,* 180–183.
96. Pukui, Haertig, and Lee, *Nānā I Ke Kumu,* vol. 1, 166–167.
97. Handy, Handy, and Pukui, *Native Planters in Old Hawaii,* 287–289.
98. C. J. Lyons, "Land Matters in Hawaii," the *Islander,* July 30, 1875.
99. Kamakau, *Works of the People of Old,* 8; Kamakau, "Ka Moolelo Hawaii: Helu 5," *Ke Au Okoa*, November 11, 1869.
100. Handy, Handy, and Pukui, *Native Planters in Old Hawaii,* 54–55, 287–288. In chapter 3, I explain in more detail how maka'āinana living ma kai traded goods and materials with those living ma uka.
101. Andrade, *Hā'ena*, 74–75; Handy, Handy, and Pukui, *Native Planters in Old Hawaii,* 321–323; Kame'eleihiwa, *Native Land and Foreign Desires,* 29–31.
102. Kamakau, *Ruling Chiefs,* 377; Pukui and Elbert, *Hawaiian Dictionary,* 334. Pō'alima was a late convention.
103. Handy, Handy, and Pukui, *Native Planters in Old Hawaii,* 288.
104. Kanahele, *Kū Kanaka, Stand Tall,* 181.
105. Pukui, Haertig, and Lee, *Nānā I Ke Kumu,* vol. 2, 160–163; Pukui, *'Ōlelo No'eau,* 17, 121.
106. I know of several places on the island of Maui where maka'āinana buried their dead next to their homes. In other places, the dead were hidden in caves and other secret places. Kamakau, *Ruling Chiefs,* 376–377; Kamakau, "Ka Moolelo Hawaii: Helu 115: No ka Noho Alii ana o Kauikeaouli Maluna o ke Aupuni, a Ua Kapaia o Kamehameha III," *Ke Au Okoa,* May 13, 1869; Kamakau, "Ka Moolelo Hawaii: Helu 116," *Ke Au Okoa,* May 20, 1869.
107. Handy, Handy, and Pukui, *Native Planters in Old Hawaii,* 47. See this source for a discussion about major land divisions.
108. Melody Kapilialoha MacKenzie, *Native Hawaiian Rights Handbook* (Honolulu: Native Hawaiian Legal Corporation, 1991), 8.

CHAPTER 3. Fluidity of Place

1. Kamakau, *Works of the People of Old,* 13; Joseph M. Poepoe, "Ka Moolelo Hawaii Kahiko," *Ka Na'i Aupuni*, February 2, 1906.
2. Peleiōhōlani spells this name two different ways. On page 32 of his translated manuscript, "Hoakalani" is written, while "Hoakaailani" is referenced on the following page. This document is housed in the Bishop Museum Archives. It is said to be a manuscript written by Peleiōhōlani and translated by Joseph M. Poepoe. S. L. Peleioholani, "The Ancient History of Hookumu-Ka-Lani=Hookumu-Ka-Honua," in *Bishop Museum Archives* (HI.L.1.3.1), 32–33.

3. Ibid., 31–33; Poepoe, "Ka Moolelo Hawaii Kahiko," *Ka Na'i Aupuni*, February 2, 1906.
4. Kamakau, "Ka Moolelo Hawaii," *Ke Au Okoa*, October 14, 1869; "Moolelo Hawaii," *Ka Hoku o Hawaii*, December 21, 1911; Poepoe, "Ka Moolelo Hawaii Kahiko," *Ka Na'i Aupuni*, February 2, 1906.
5. Kamakau, *Tales and Traditions of the People of Old*, 128; *Mauiloa: A Collection of Mele for the Islands of Maui, Kaho'olawe, Lāna'i, Moloka'i*, ed. Ku'ulei Higashi Kanahele, trans. Pualani Kanaka'ole Kanahele (2005), 7, Poepoe.
6. R. Kawena Johnson, "Hawaiian 261" (University of Hawai'i, 2001), 305; Kanahele, *Mauiloa*, 7.
7. Beckwith, *Hawaiian Mythology*, 305; Johnson, "Hawaiian 261," 304; Kamakau, *Tales and Traditions of the People of Old*, 129.
8. Beckwith, *Hawaiian Mythology*, 305; Johnson, "Hawaiian 261," 304; Kamakau, *Tales and Traditions of the People of Old*, 129; Elspeth P. Sterling and Native Hawaiian Culture and Arts Program, *Sites of Maui* (Honolulu: Bishop Museum Press, 1998), 2.
9. Beckwith, *Hawaiian Mythology*, 305; Johnson, "Hawaiian 261," 304; Kamakau, *Tales and Traditions of the People of Old*, 129; State of Hawai'i Division of Forestry and Wildlife and State of Hawai'i Department of Land and Natural Resources, *Wao Akua: Sacred Source of Life* (Honolulu: Department of Land and Natural Resources, 2003), 3.
10. Peleioholani, "Ancient History of Hookumu-Ka-Lani," 33.
11. The island of Maui was named after the demigod Māui. Some people insist that the correct spelling is Māui, while others do not use the kahakō because native speakers do not elongate the "a" sound in their pronunciation of the island name. In *Place Names of Hawaii*, it is acknowledged that the island is named after the demigod Māui; however, it is spelled without the kahakō. Pukui, Elbert, and Mookini, *Place Names of Hawaii*, 148.
12. Kamakau, *Tales and Traditions of the People of Old*, 128–129.
13. Johnson, "Hawaiian 261," 304, 325.
14. Yi-fu Tuan, "Language and the Making of Place: A Narrative-Descriptive Approach," *Annals of the Association of American Geographers* 81:4 (1991): 686.
15. Ka lewa was also known as "ka hookui" and "ka halawai." Malo, *Hawaiian Antiquities*, 9. Emerson, who translated Malo's work from 'ōlelo makuahine into English, contends that "*Hookui* is undoubtedly that part of the vault of heaven, the zenith, where the sweeping curves of heaven's arches meet; the *hala-wai* was probably the line of junction between the *kukulu*, walls or pillars on which rested the celestial dome, and the plane of the earth." Emerson continues, "*Ka halawai* is probably applied to the horizon, the line where the walls of heaven join the plain of the earth" (ibid., 11nn5–6). Kamakau's analysis of the same topics reveals

the following: "Equidistant from the sky downward and the earth upward—was called *ka ho'oku'i*, the juncture, or *ka ho'ohalawai*, the meeting." Kamakau, *Works of the People of Old,* 6; Kamakau, "Ka Moolelo Hawaii: Helu 4," *Ke Au Okoa,* November 4, 1869.

16. Kamakau, "Ka Moolelo Hawaii: Helu 4" *Ke Au Okoa,* November 4, 1869; Kamakau, *Works of the People of Old,* 6; Malo, *Hawaiian Antiquities,* 10. In *The Works of the People of Old*, Pukui translates "Kamaku'ialewa" to mean "the circle of air that surrounds the earth [the atmosphere]" (6). In the *Hawaiian Dictionary* by Pukui and Elbert, "kama-kū-i-kahi-lewa" is defined as "the sky just below the zenith" (125).
17. Pukui, *'Ōlelo No'eau,* 7. Hereafter cited in text as "ON."
18. Kamakau, *Ka Po'e Kahiko,* 8; Kamakau, "Ka Moolelo Hawaii: Helu 47," *Ke Au Okoa,* November 3, 1870; Malo, *Hawaiian Antiquities,* 7, 248.
19. Kamakau, *Ka Po'e Kahiko,* 8; Kamakau, "Ka Moolelo Hawaii: Helu 47," *Ke Au Okoa,* November 3, 1870.
20. According to Kamakau, the ruling chiefs of Hilo, Hawai'i, were cousins to some Maui chiefs; therefore, Kamalālāwalu only chose to declare war on the chiefs of Kohala, Kona, and Ka'ū. Kamakau, *Ruling Chiefs*, 55.
21. Ibid., 57–58.
22. Pukui, Haertig, and Lee, *Nānā I Ke Kumu,* vol. 2, 272–273.
23. Kamakau, "Ka Moolelo Hawaii: Helu 5," *Ke Au Okoa*, November 11, 1869; Kamakau, *Works of the People of Old,* 8.
24. Malo, *Hawaiian Antiquities,* 17. According to Emerson's footnotes in David Malo's *Hawaiian Antiquities,* "a *kua-lono* was a broad plateau between two valleys, while a *kua-lapa* was a narrow ridge." Ibid., 18. Pukui and Elbert define "*kualono*" as a "region near the mountaintop, ridge," "kualapa" as "ridge," and "kuamauna" as "mountaintop." Pukui and Elbert, *Hawaiian Dictionary,* 170; Kamakau, "Ka Moolelo Hawaii: Helu 5," *Ke Au Okoa*, November 11, 1869.
25. According to Kamakau, small trees grew in the wao nahele region. Kamakau, "Ka Moolelo Hawaii: Helu 5," *Ke Au Okoa*, November 11, 1869; Kamakau, *Works of the People of Old,* 9. Contradictorily, Malo states that large, forest-sized trees populated this zone. Malo, *Hawaiian Antiquities,* 17.
26. Tall trees dominated the wao lipo. Kamakau, "Ka Moolelo Hawaii: Helu 5," *Ke Au Okoa*, November 11, 1869; Kamakau, *Works of the People of Old,* 9.
27. Large, forest-sized trees were found here. Kamakau, "Ka Moolelo Hawaii: Helu 5," *Ke Au Okoa*, November 11, 1869; Malo, *Hawaiian Antiquities,* 17.
28. Koa and 'ōhi'a trees dominate the canopy here. Division of Forestry and Wildlife, *Wao Akua,* 11; Kamakau, *Works of the People of Old,* 9; Kamakau, "Ka Moolelo Hawaii: Helu 5," *Ke Au Okoa*, November 11, 1869; Malo, *Hawaiian Antiquities,* 17.

29. The wao akua was found below the wao maʻukele. Division of Forestry and Wildlife, *Wao Akua,* xiv, 8–14; Kamakau, *Works of the People of Old,* 9; Kamakau, "Ka Moolelo Hawaii: Helu 5," *Ke Au Okoa*, November 11, 1869; Ka Ohu Haaheo i na Kuahiwi Ekolu, "Ka Waiu Ame Ka Meli," *Ka Hoku o Hawaii,* June 17, 1920.
30. Division of Forestry and Wildlife, *Wao Akua,* xiv; Kamakau, "Ka Moolelo Hawaii: Helu 5," *Ke Au Okoa*, November 11, 1869; Kamakau, *Works of the People of Old,* 9; Ka Ohu Haaheo i na Kuahiwi Ekolu, "Ka Waiu Ame Ka Meli," *Ka Hoku o Hawaii,* June 17, 1920.
31. Kamakau, "Ka Moolelo Hawaii: Helu 5," *Ke Au Okoa*, November 11, 1869; Kamakau, *Works of the People of Old,* 7; Malo, *Hawaiian Antiquities,* 16. According to Handy, Handy, and Pukui, "*honua* is mass and *ʻaina* [*sic*] is surface." They further explain that honua is commonly used in reference to plain areas, but likewise refers to the earth or land in general, while ʻāina means land in the sense of homeland and birthplace. Handy, Handy, and Pukui, *Native Planters in Old Hawaii,* 43–44.
32. "Moku kele i ka waʻa" was often abbreviated as "moku kele." Pukui and Elbert, *Hawaiian Dictionary*, 143, 252.
33. Kamakau, "Ka Moolelo Hawaii: Helu 5," *Ke Au Okoa*, November 11, 1869; Kamakau, *Works of the People of Old,* 7; Malo, *Hawaiian Antiquities,* 16; Ka Ohu Haaheo i na Kuahiwi Ekolu, "Ka Waiu Ame Ka Meli," *Ka Hoku o Hawaii,* June 17, 1920.
34. Beckwith, *Hawaiian Mythology,* 383; Handy, Handy, and Pukui, *Native Planters in Old Hawaii,* 491.
35. Handy, Handy, and Pukui, *Native Planters in Old Hawaii,* 47.
36. Kamakau, "Ka Moolelo Hawaii: Helu 5," *Ke Au Okoa*, November 11, 1869; Kamakau, *Works of the People of Old,* 7.
37. Malo, *Hawaiian Antiquities,* 16.
38. Kamakau, "Ka Moolelo Hawaii: Helu 5," *Ke Au Okoa*, November 11, 1869; Kamakau, *Works of the People of Old,* 7. This is consistent with a June 17, 1920, newspaper article, "Ka Waiu Ame Ka Meli," appearing in *Ka Hoku o Hawaii,* written by someone using the pen name, "Ka Ohu Haaheo i na Kuahiwi Ekolu."
39. *Buke Mahele* is the book that recorded the Māhele, the division of lands between Kamehameha III, the aliʻi, and the konohiki that led to the privatization of lands in ka pae ʻāina Hawaiʻi.
40. W. D. Alexander, "Appendix 1," in *A Brief History of Land Titles in the Hawaiian Kingdom* (Honolulu: P. C. Advertsier Co., 1882), 4.
41. Pukui and Elbert, *Hawaiian Dictionary,* 121. This may suggest that the way in which land terms were defined may have changed over time.
42. Mokuleia writes, "O na mokupuni, oia na mea nui, ua mahele ia oia i mau apana, a o kela mau apana, i mahele ia ua ka-pa ia he moku o loko, e like me Kohala i

Hawaii, o Hawaii ka mokupuni, a o Kohala hoi ka moku o loko o Hawaii, a pela aku no. . . . Eia he wahi mahele hou, o ka okana, a he kalana kekahi inoa." S. Mokuleia, "Kekahi mau mea a ka poe kahiko i mahele ai o Hawaii nei i ka wa kahiko," *Ka Nupepa Kuokoa*, March 7, 1868. Andrews' definition of kalana: "The name of a division of an island next less than moku, and SYN with okana in some places." further affirms Malo's assertion. Lorrin Andrews, *A Dictionary of the Hawaiian Language* (Waipahu: Island Heritage Publishing, 2003), 247.

43. R. D. King, "Districts in the Hawaiian Islands," in J. W. Coulter, comp., *A Gazetteer of the Territory of Hawaii* (University of Hawai'i Research Pub. No. 11, 1935), 215.
44. Pukui and Elbert, *Hawaiian Dictionary*, 9.
45. C. J. Lyons, "Land Matters in Hawaii—No. 2," *The Islander*, July 9, 1875, 111.
46. Donovan Preza, *The Empirical Writes Back: Re-examining Hawaiian Dispossession Resulting from the Māhele of 1848* (Honolulu: University of Hawai'i at Mānoa, 2010), 61–62.
47. Alexander, *Brief History of Land Titles*, 4; Handy, Handy, and Pukui, *Native Planters in Old Hawaii*, 48-49; Kamakau, *Ka Po'e Kahiko*, 39; Lyons, "Land Matters in Hawaii," July 9, 1875, 111.
48. Kapulani Landgraf, *Nā Wahi Kapu o Maui* (Honolulu: 'Ai Pōhaku Press, 2003), 108.
49. Kamakau, *Works of the People of Old*, 7; Kamakau, "Ka Moolelo Hawaii: Helu 5," *Ke Au Okoa*, November 11, 1869; Lyons, "Land Matters in Hawaii," 118–119; Malo, *Hawaiian Antiquities*, 16.
50. Pukui and Elbert, *Hawaiian Dictionary*, 97.
51. Lyons, "Land Matters in Hawaii," 118–119.
52. A kō'ele was a small piece of land farmed by the maka'āinana of an ahupua'a for an ali'i. Malo, *Hawaiian Antiquities*, 18.
53. Emerson explains in the footnotes of Malo's *Hawaiian Antiquities*, "*Haku-one* was the small piece of land under cultivation by the peasant which the *konohiki* seized for his own use, though the peasant had to continue its cultivation. A peasant, for instance, had six *taro* patches; the *alii* appropriated the best one for himself, and that was called *koele*. The *konohiki*, or *haku-aina*, took another for himself and that was called *haku-one*." Malo, *Hawaiian Antiquities*, 18n2.
54. A kuakua was wider than a kuauna. Both were embankments dividing one lo'i from another. Ibid., n3.
55. Kamakau, "Ka Moolelo Hawaii: Helu 5," *Ke Au Okoa*, November 11, 1869; Kamakau, *Works of the People of Old*, 7–8; Malo, *Hawaiian Antiquities*, 16.
56. Kanahele, *Kū Kanaka, Stand Tall*, 188–194.
57. Kamakau, *Works of the People of Old*, 3–4; Kamakau "Ka Moolelo Hawaii: Helu 4," *Ke Au Okoa*, November 4, 1869.

58. Handy, Handy, and Pukui, *Native Planters in Old Hawaii,* 54–55.
59. E. Hauʻofa, "Our Sea of Islands," *Contemporary Pacific: A Journal of Island Affairs* 6:1 (1994) 148–161.
60. Kamakau, "Ka Poʻe Kahiko," 74; Kamakau, "Ka Moolelo Hawaii: Helu 26," *Ke Au Okoa*, April 14, 1870. The ocean was a sacred place; seawater was used in religious ceremonies for purification purposes. Many gods occupied the ocean in the form of sharks, eels, and other marine life. The sea was also sacred in that it served as a burial ground for some humans, especially those whose ʻaumākua were physical manifestations of Kanaloa, the god of the deep ocean. Even fishing spots were sacred, as they were off limits to fishing during certain times of the year. Kanahele, *Kū Kanaka, Stand Tall,* 192. For the most current resources related to Kanaka navigation, please refer to the Polynesian Voyaging Society's website: http://hokulea.org/education/.
61. For a more in-depth discussion about performance cartographies, please refer to: Jay T. Johnson, Renee Pualani Louis, and Albertus Hadi Pramono, "Facing the Future: Encouraging Critical Cartographic Literacies in Indigenous Communities," *ACME: An International E-Journal for Critical Geographies* 4:1 (2006): 80–98; Woodward and Lewis, *History of Cartography.*
62. Sterling and Native Hawaiian Culture and Arts Program, *Sites of Maui,* 9.
63. Personal communication with caretakers of Piʻilani Hale heiau.
64. Malo, *Hawaiian Antiquities,* 25–26.
65. Personal family knowledge; Andrade, *Hāʻena,* 67.
66. Isabella Aiona Abbott, *Lāʻau Hawaiʻi: Traditional Hawaiian Uses of Plants* (Honolulu: Bishop Museum Press, 1992), 45.
67. Division of Forestry and Wildlife, *Wao Akua,* xiv–xv.
68. The eight channels referred to here are those between Lahaina and Molokaʻi, Molokaʻi and Lānaʻi, Lānaʻi and Kaulakoʻi, Lānaʻi and Kahoʻolawe, Kahoʻolawe and Honuaʻula, Olualu [sic] and Kahoʻolawe, Lahaina and Lānaʻi, and Kahakuloa and Molokaʻi. Sterling and Native Hawaiian Culture and Arts Program, *Sites of Maui,* 9; Pukui, *ʻŌlelo Noʻeau,* 243.
69. Personal communication, June 6, 2013. There are actually nine channels between the eight major islands in ka pae moku Hawaiʻi.
70. Interestingly, "ʻākau" and "hema" not only refer to the cardinal points north and south but also to the right and left sides of a person's body. Even a circuit of the islands by aliʻi incorporated these two terminologies. Should an aliʻi choose to make a counterclockwise circuit of the island, keeping his hema arm toward land, such a circuit was known as "kaʻa lalo kūkulu hema." Conversely, traversing the island with one's ʻākau arm to the land was known as "kaʻa kūkulu ʻākau"

and symbolized maintenance of control over the ʻāina. Because the direction of travel made by an aliʻi had cultural significance, a circuit of the islands was a way of mapping a chief's landholdings and inscribing place with meaning. Kamakau, *Works of the People of Old,* 3–4; Kamakau, "Ka Moolelo Hawaii: Helu 4," *Ke Au Okoa*, November 4, 1869. Kāohilani is a poetic name for left or the left hand. Pukui and Elbert, *Hawaiian Dictionary,* 130.

71. Handy, Handy, and Pukui, *Native Planters in Old Hawaii,* 23, 340; Kanahele, *Kū Kanaka, Stand Tall,* 47–48.
72. Sterling and Native Hawaiian Culture and Arts Program, *Sites of Maui,* 3. According to King, Kahoʻolawe was grouped with the Koʻolau District of Maui; however, Sterling says Kahoʻolawe was a part of the Honuaʻula District; Nanea K. Armstrong, "Moving Time and Land: Cultural Memories through Inscribed Landscapes" (M.A. thesis, University of Hawaiʻi at Mānoa, 2003).
73. http://www.usgs.gov/ecosystems/pierc/research/maui.html (retrieved May 20, 2013). See also John L. Culliney, *Islands in a Far Sea: The Fate of Nature in Hawaiʻi* (Honolulu: University of Hawaiʻi Press, 2006), 17. Culliney contends that for a period of approximately 300,000 years, Oʻahu was also connected to Maui Nui via a land bridge.
74. Stephen B. Jones, "Geography and Politics in the Hawaiian Islands," *Geographical Review* 28:2 (April 1938), 205.
75. Kamakau, "Ka Moolelo o Kamehameha I: Helu 11," *Ka Nupepa Kuokoa*, January 19, 1867.
76. *Native Planters* spells this place name "Honokahau" and "Honokohau," both without kahakō. In fact, kahakō are absent from the book, but ʻokina are utilized. Handy, Handy, and Pukui, *Native Planters in Old Hawaii,* 272, 494. In *Place Names of Hawaii,* "Honokāhau" and "Honokōhau" are given as variations. To be consistent, I have chosen to use "Honokōhau," as it is the most common pronunciation heard today on Maui. Pukui, Elbert, and Mookini, *Place Names of Hawaiʻi,* 49.
77. Handy, Handy, and Pukui, *Native Planters in Old Hawaii,* 272.
78. Jones, "Geography and Politics in the Hawaiian Islands," 206; Lyons, "Land Matters in Hawaii," 119.
79. *Buke Mahele.*
80. According to Stephen B. Jones, the isthmus was not incorporated into a district until 1859, when four districts were created. "Geography and Politics in the Hawaiian Islands," 206; King, "Districts in the Hawaiian Islands," 216.
81. King, "Districts in the Hawaiian Islands," 216.
82. Ibid., 216–217.
83. Ibid., 218.

84. Ibid., 219–223.
85. Kamakau, *Works of the People of Old,* 7; Kamakau, "Ka Moolelo Hawaii: Helu 5," *Ke Au Okoa*, November 11, 1869.
86. Alexander, *Brief History of Land Titles,* 6.
87. Kamakau, *Works of the People of Old,* 3–6; Kamakau, "Ka Moolelo Hawaii: Helu 4," *Ke Au Okoa*, November 4, 1869.
88. Kanahele, *Kū Kanaka, Stand Tall* 180. For further discussion, please see Handy, Handy, and Pukui, *Native Planters in Old Hawaii.*
89. Division of Forestry and Wildlife, *Wao Akua,* 8–14; Johnson, Louis, and Pramono, "Facing the Future"; Woodward and Lewis, *History of Cartography,* 87.
90. J. H. Wise, "The History of Land Ownership in Hawaii," ed. Kamehameha Schools (Rutland, VT, & Tokyo: Charles E. Tuttle Co., 1965), 84–85.
91. From the 1850s through the 1920s, Mākena was the second largest port on the island of Maui, transporting people, sending products, and maintaining a cultural and linguistic link between the people of Maui with the rest of Hawaiʻi. Armstrong, "Moving Time and Land," 35. Today, the interaction of places still occurs—now, however, by way of air travel. No matter the mode of transportation, such interisland travel continues to allow for the exchange of resources between the islands. And, the tradition of gift giving remains embedded in our Hawaiian culture.
92. Andrade, *Hāʻena*, 99.
93. When trying to purchase a share in a hui, I learned very quickly that securing a mortgage would be difficult. I could not use the hui property as collateral.
94. Aliʻi nui would try to mitigate some of these consequences in order to reduce the burden of the kapu placed on the makaʻāinana. Aliʻi nui, for example, often did not go out during the daylight hours, lest their shadows would make the ground they walked upon sacred. "Na Lei o Hawaii," *Ka Nupepa Kuokoa,* February 24, 1883; "Na Papa Alii o Hawaii," *Ka Nupepa Kuokoa*, March 10, 1883; Kamakau, "Ka Moolelo Hawaii: Helu 46," *Ke Au Okoa*, October 27, 1870; Kanahele, *Kū Kanaka, Stand Tall,* 196.
95. Pukui and Elbert, *Hawaiian Dictionary,* 437.
96. Tuan, "Language and the Making of Place," 686.

CHAPTER 4: Performance Cartographies

1. In their book, Woodward and Lewis identify three categories of cartographic representations common amongst the traditional societies of Africa, America, Arctic, Australia, and Pacific Islanders: cognitive, performance, and material cartographies. They contend that cognitive cartography is used "to denote physical artifacts recording how people perceive places." Performance cartography "fulfills

the function of a map" and "may take the form of a nonmaterial oral, visual, or kinesthetic social act, such as a gesture, ritual, chant, procession, dance, poem, story, or other means of expression or communication whose primary purpose is to define or explain spatial knowledge or practice" (4). In reference to material cartography, Woodward and Lewis make the following statement: "A spatial representation may also be a permanent or at least nonephemeral record created or placed in situ, as in rock art, maps posted as signs, or maps embodied in shrines or buildings. Or the representation may take the form of a mobile, portable, archivable record" (5). Of all of these cartographic categories, performance cartography was most commonly used by ancestral Kānaka. Therefore, the examples in this book with reference to ancestral Kanaka cartographic methods will be derived largely from this category. Woodward and Lewis, *History of Cartography,* 1–5; Pukui, *ʻŌlelo Noʻeau.* For a more in-depth discussion about performance cartographies, please refer to Johnson, Louis, and Pramono, "Facing the Future"; Woodward and Lewis, *History of Cartography,* 81–98.

2. Woodward and Lewis, *History of Cartography,* 4.
3. Jan Kelly, "Maori Maps," *Cartographica* 32:6 (1999), 1.
4. Kānaka did not create written maps in ancestral times. Thus, the terms "cartography" and "cartographic" have been placed in quotes the first time they appear in this book to indicate that Kānaka had our own approaches for "mapping" our environment.
5. Gregory Cajete, *Native Science: Natural Laws of Interdependence* (Santa Fe, NM: Clear Light Publishers, 2000), 182–183.
6. Edward S. Casey, *Remembering: A Phenomenological Study* (Bloomington: Indiana University Press, 2000), 182.
7. Cajete, *Native Science,* 205.
8. For lack of a more culturally appropriate term, "map" will be used throughout this book to approximate Kanaka performance cartography techniques. "Map" appears in quotes in the first usage of the term because ancestral Kānaka did not have maps as we know them today. Rather than drawing maps on paper, Kānaka used performance to record location, proximity, and direction.
9. The word "pōhaku" means "rock." Therefore, it has the connotation of being the foundation of one's existence. Pukui and Elbert, *Hawaiian Dictionary*, 334.
10. The numerous mele wahi recorded in pana Hawaiʻi newspapers of the nineteenth and early twentieth centuries as well as in archival collections are a testament to the importance of mele wahi pana.
11. "Ua Noho Au a Kupa," composed by Edward Nainoa in 1890.
12. I have selected a sampling of mele related to the island of Maui to demonstrate how mele may serve as performative cartographies and Kanaka geographies. I have chosen mele that I feel best exemplify the points I am making. Every

attempt has been made to select mele that the reader might have heard before or that Hawai'i radio stations might commonly play. The analysis of these mele provides a very brief overview of how the mele may be interpreted; however, it is beyond the scope of this book to discuss all of the possible layers of meaning concealed in the mele by the composer.

13. T. Davis, T. O'Regan, and J. Wilson, *Nga Tohu Pumahara: The Survey Pegs of the Past* (Wellington: New Zealand Geographic Board, 1990), 7.
14. The lyrics for this mele appear here as they are recorded in Kimo 'Alama Keaulana's mele collection housed at the Bishop Museum. The English translation is mine. Kimo 'Alama Keaulana, "Koali," in Mele Collection of Kimo Alama Keaulana (Honolulu).
15. Te Rangi Hiroa (Sir Peter H. Buck), *Arts and Crafts of Hawaii* (Honolulu: Bishop Museum Press, 2003), 570–571.
16. The lyrics are recorded here as they appear in Carol Wilcox et al., *He Mele Aloha: A Hawaiian Songbook* (Honolulu: 'Oli'oli Productions, 2004), 107. The English translation is mine.
17. This is Ka'upena Wong's translation for these lines as it appears in Nona Beamer's *Nā Mele Hula,* vol. 2: *Hawaiian Hula Rituals and Chants,* 66–67.
18. Kāne is sometimes credited as being a po'olua (double paternity) along with Wākea of Maui(loa). Fornander, *Fornander Collection,* vol. 4, 12, 15; Kanahele, *Mauiloa*, 6.
19. Kanahele, *Mauiloa,* 6–7.
20. Examples of kanikau composed for ali'i may be found in volume 6 of the *Fornander Collection.*
21. Fornander, *Fornander Collection,* vol. 6, 427–429.
22. There are variations of spelling. Rubellite Kawena Johnson spells Maui's former name "Ihikapulaumaewa" on pages 31–32 in *Essays in Hawaiian Literature,* while Kamakau in "Ka Moolelo o Hawaii: Helu 2," appearing on October 21, 1869 in the 'ōlelo makuahine newpaper *Ke Au Okoa,* spells it "Ihikapalaumaewa." See also Kamakau, *Tales and Traditions,* 129.
23. Sterling and Native Hawaiian Culture and Arts Program, *Sites of Maui,* 2. "Kūlua" means "twins." The island is often called a double island because of its geography.
24. Beckwith, *Hawaiian Mythology,* 305; Johnson, "Hawaiian 261," 304; Kamakau, *Tales and Traditions of the People of Old,* 129; Elspeth P. Sterling and Native Hawaiian Culture and Arts Program, *Sites of Maui* (Honolulu: Bishop Museum Press, 1998), 2.
25. Davis, O'Regan, and Wilson, *Nga Tohu,* 7.
26. Kanahele, *Kū Kanaka, Stand Tall,* 184.
27. Pukui, Elbert, and Mookini, *Place Names of Hawaii,* x.

28. Although the island is named after the akua Māui, according to *Place Names of Hawaii* the island name, Maui, is spelled without a kahakō, even though the akua for which it is named *is* spelled with a kahakō over the "a." I have yet to hear a mānaleo pronounce the island name with a kahakō. Ibid., 148.
29. Fornander, *Fornander Collection,* vol. 5, 536.
30. Liliuokalani, *Kumulipo,* 75.
31. Many variations exist for this place name. According to Fornander, Māui's parents resided in Makaliua (*Fornander Collection,* vol. 5, 536). According to Martha Beckwith, the spelling of this place name is "Makalia" (*Hawaiian Mythology,* 230). In *Place Names of Hawaii,* a place fitting this same geographic location is known as "Makalina" (Pukui, Elbert, and Mookini, *Place Names of Hawaii,* 141).
32. Fornander, *Fornander Collection,* vol. 5, 536–539.
33. Kamakau, *Works of the People of Old,* 116–117. According to Beckwith, the waimea tree was the source of fire (*Hawaiian Mythology,* 229–230).
34. Beckwith, *Hawaiian Mythology,* 230; Puaaloa, "Ka Moolelo O Maui," *Ka Nupepa Kuokoa,* July 4, 1863.
35. Also known as Paeloko. Beckwith, *Hawaiian Mythology,* 231.
36. Fornander, *Fornander Collection,* vol. 5, 538–539.
37. W. D. Westervelt, *Legends of Maui: A Demi-God of Polynesia and of His Mother Hina* (Melbourne: G. Robertson & Co., 1913), 31; W. D. Westervelt, *Hawaiian Historical Legends* (New York: Fleming H. Revell Co., 1923), 15.
38. Beckwith, *Hawaiian Mythology,* 230.
39. Pukui, *ʻŌlelo Noʻeau,* 170. Hereafter cited in text as "ON."
40. Tuan, "Language and the Making of Place," 686.
41. Here the words "lilo" and "loa" are used to emphasize distance. Lilo refers to something that is far away and distant, perhaps to the point that it is out of sight. Loa qualifies lilo, again implying an even greater distance. Notice how lilo and loa are used in conjunction with directionals to illustrate the remoteness of place and/or time.
42. Malo, *Hawaiian Antiquities,* 12.
43. Ibid., 15.
44. "Kūkunu" is another word for "kūkulu." Pukui and Elbert, *Hawaiian Dictionary,* 178.
45. North corresponds to the placement of the star Kahōkūpaʻa, in the sky. Kahōkūpaʻa is a Kuanalio or Lio star (North Pole star). The term "kūkulu ʻākau" is also used to refer to the north. "Kūkulu" means pillar, post, border, edge, or horizon. From a Kanaka perspective, then, these cardinal points are the borders of the earth, conceptually propping up the sky. There are two types of kūkulu: one, like the horizon, is seen, while the second type is invisible. Ulunui,

Uliuli, Melemele, Hakalau'ai, and Kuanalio were alternate names for the north. Kamakau, *Works of the People of Old,* 3–5; Kamakau, "Ka Moolelo Hawaii: Helu 4," *Ke Au Okoa,* November 4, 1869. According to Malo, the north is also referred to as "luna" or "i luna," while the south is called "lalo" (*Hawaiian Antiquities,* 9). 'Elekū is a name used by priests for the north. Pukui and Elbert, *Hawaiian Dictionary,* 40. Ho'olua was a wind from the north. Malo, *Hawaiian Antiquities,* 14.

46. South, also known as kūkulu hema, Kuanalipo, and Kuanalepo, makes reference to the direction toward the lipo and lewa, the depths of the deep blue ocean. Stars in the South Pole direction are known as Lipo or Lioliowawau. Kamakau, *Works of the People of Old,* 4; Kamakau, "Ka Moolelo Hawaii: Helu 4," *Ke Au Okoa,* November 4, 1869. The south is associated with Kona winds. Malo, *Hawaiian Antiquities,* 14. 'Elemoe is a name used by priests to refer to the south. Pukui and Elbert, *Hawaiian Dictionary,* 40.
47. The rising of the sun marked the eastern cardinal point, kūkulu hikina. Kānaka associated the rising of the sun with the coming of life and blessings. Names such as Kahikina, Kalāhikiola, Kalāikamauliola, and Kalāikalanaola all refer to the arrival of the sun and its ability to give life to the land. On the island of Maui, east was also known as Na'e because this wind came from an easterly direction. Kamakau, *Works of the People of Old,* 4; Kamakau, "Ka Moolelo Hawaii: Helu 4," *Ke Au Okoa,* November 4, 1869; Malo, *Hawaiian Antiquities,* 10. 'Elelani was a priestly reference to the east. Pukui and Elbert, *Hawaiian Dictionary,* 40.
48. The west was known by kāhuna as kūkulu lā kau. Komohana means "entering into." Because this meaning was considered an ill omen, kāhuna referred to the west as ka lā kau or kaulana, the place where the sun set, rather than kūkulu komohana. Nā kukuna o ka lā kau, the rays of the setting sun, was another other reference for the west. Kamakau, *Works of the People of Old,* 4; Kamakau, "Ka Moolelo Hawaii: Helu 4," *Ke Au Okoa,* November 4, 1869. 'Elehonua and 'eleiāhonua were priestly names for the west. Kapa kai sometimes referred to the west. Pukui and Elbert, *Hawaiian Dictionary,* 40, 131. See also Malo, *Hawaiian Antiquities,* 9–10.
49. Kamakau, *Works of the People of Old,* 3–5; Kamakau, "Ka Moolelo Hawaii: Helu 4," *Ke Au Okoa,* November 4, 1869; Malo, *Hawaiian Antiquities,* 9–11.
50. Tuan, "Language and the Making of Place," 695.
51. Deborah Hill, "Distinguishing the Notion 'Place' in Oceanic Language," in Martin Pütz and René Dirven, eds., *The Construal of Space in Language and Thought* (Berlin: Mouton de Gruyter, 1996).
52. "Lā" is sometimes spelled "ala."
53. Judd, *Hawaiian Language and Hawaiian-English Dictionary,* 29, 37; Samuel H. Elbert and Mary Kawena Pukui, *Hawaiian Grammar* (Honolulu: University of Hawai'i Press, 1979), 60–61, 113.

54. Kanahele, *Kū Kanaka, Stand Tall,* 184.
55. Because winds were often accompanied by rain, sometimes a single name was applied to both the wind and rain of a place.
56. One resource for the names of the winds and rains of parts of Eastern Maui, particularly Kaupō, is found in *Nupepa Kuokoa,* June 15, 1922. This newspaper article, along with other articles written by Thomas K. Maunupau about Kaupō, was published under the following title: *Huakai Makaikai a Kaupo, Maui: A Visit to Kaupō, Maui: As Published in Ka Nupepa Kuokoa, June 1, 1922–March 15, 1923* (Honolulu: Bishop Museum Press, 1998).
57. Because many contradictory versions of this mele exist, I have chosen to use the text as written by Kānepu'u in "Kaahele ma Molokai: Helu 5" in *Ke Au Okoa,* October 17, 1867. Only the first half of the mele, the portion pertaining to Maui, has been included, while those in the latter half, those relating to Moloka'i, have been omitted. Variations appearing in *Mauiloa* and *Sites of Maui* have been noted (page 7). Another version can be found in Moses K. Nakuina, *Moolelo Hawaii o Pakaa a me Ku-a-Pakaa* (Honolulu, 1902).
58. Fornander, *Fornander Collection,* vol. 5, 72–77.
59. Nāulu is characterized as a strong but short-lived windstorm accompanied by rain. Malo, *Hawaiian Antiquities,* 14.
60. Hau is a land breeze, often blowing from the mountains down toward the sea. Ibid., 14–15.
61. *Mauiloa* spells this place name Liliko'i (page 15), while *Sites of Maui* calls this place Lilikoa (page 7). Lilikoi, Liliko'i, and Lilikoa are all absent from *Place Names of Hawaii.*
62. In *Ke Au Okoa,* (October 17, 1867) this place name is spelled Ukumehama, while *Sites of Maui* (page 7) and *Mauiloa* (page 15) both call this place Ukumehame.
63. *Mauiloa* (page 16) spells it "Laiki"; however, *Sites of Maui* (page 7) spells it "Iaiki." This discrepancy may be due to the fact that a lower case "l" and an upper case "I" are sometimes identical depending on the font utilized.
64. Malo states that the 'A'a wind at Lahaina travels from the sea inland on the leeward side of an island (Malo, *Hawaiian Antiquities,* 14). Perhaps the Ma'a'a wind enumerated in this mele is another version of 'A'a (Pukui and Elbert, *Hawaiian Dictionary,* 1).
65. In *Mauiloa,* Paalaa is spelled Pā'ala (page 16).
66. Malo says that the 'A'a wind that blows from the sea toward the land has other names at different places. In Lahaina, Maui, it is known as 'A'a, but he offers the names 'Eka, Kaiāulu, and Inuwai as alternatives for other places. It is not known if the Kaiāulu mentioned here is the same as the 'A'a wind of Lahaina. Malo, *Hawaiian Antiquities,* 14.
67. While the word "'āina" is often translated to mean "land," the root of this word

is "'ai," meaning "to eat." In addition to referring to the land, 'ai emphasizes that people do not only *live* on the land but also *eat* off of its resources.

CHAPTER 5: Ancestral Sense Abilities

1. Pukui and Elbert, *Hawaiian Dictionary,* 96.
2. Pukui, Haertig, and Lee, *Nānā I Ke Kumu,* vol. 2, 270.
3. Pukui, *'Ōlelo No'eau,* 20. Hereafter cited in text as "ON."
4. Kamakau, *Ruling Chiefs,* 132.
5. Pukui, Haertig, and Lee, *Nānā I Ke Kumu,* vol. 2, 171–185.
6. It was believed that such inoa pō were destined for the child and must be given, or else sickness or death might result. Hō'ailona are sources of knowledge for Kānaka. Ibid., 169–185, 269–283.
7. Lisa Schattenburg-Raymond, "Nā Waiho'olu'u O Ke Ānuenue," in *Ka 'Aha Hula 'O Hālauaola* (Maui Arts and Culture Center and Maui Community College, 2005).
8. It was believed that poi should be made only during the daylight hours in front of the house so that any passerby could plainly see the kalo being pounded. This prevented others from thinking that the family was stingy and did not want to share their poi. Personal communication, Edward Kaanaana, 1996–2006.
9. Personal Communication, Maryann Nākoa Barros; Hospice Hawai'i, *Volunteer Handbook: Resources, Procedures and Guidelines* (Honolulu: Hospice Volunteer Services, 2011).
10. Andrews, *Dictionary of the Hawaiian Language,* 408.
11. Pukui and Elbert, *Hawaiian Dictionary,* 257.
12. Pukui, Haertig, and Lee, *Nānā I Ke Kumu,* vol. 1, 155.
13. Pukui and Elbert, *Hawaiian Dictionary,* 104.
14. Kanahele, *Kū Kanaka, Stand Tall,* 182.
15. Personal communication, Ned Nākoa, 1974–1990. My pure Kanaka grandfather taught me many important Hawaiian cultural practices. Even after his death, I continue to remember the lessons he instilled in me.
16. Mere Roberts, "Indigenous Knowledge and Western Science: Perspectives from the Pacific," *Royal Society of New Zealand: Miscellaneous Series* 50 (1996): 62.
17. Different times within the morning hours were noted. "Pawa" is a predawn time. "Alaula" and "mahikina lā" refer to the light of early dawn and the crack of dawn, respectively. The approach of dawn is known as "ka pili o ka wana'ao." "Poni li'ulā" is an early glimmer of dawn, while "wana" is a streak of light associated with dawn. "Kāhe'a" is another name for red streaks that appear at dawn. "Pohā kea" is the time that the sun bursts forth at dawn, as is "mali'o." "Kaiao" and "wana'ao"

are terms that refer to dawn. "'Ōmaka" refers to the rising of the sun. "Wehe a'ela ka 'Iao" figuratively refers to the breaking of dawn. "Wehena" is the opening or appearance of the sunlight. "Ao" refers to both dawn and daylight. "Pō iki" alludes to the transition between night and day and refers to the early morning hours. "Kakahiaka nui" likewise makes reference to the early morning hours. "'Oko'a" is early morning, but when used in the phrase "i ka lā 'oko'a," it alludes to being in broad daylight. Pukui and Elbert, *Hawaiian Dictionary.*

18. "Lolokū" and "na'uā" (also known as "na'uwā") are two words that refer to noon. The phrase "ke lolokū aku nei" means it is noontime and the sun is overhead. Ibid., 211, 263.
19. "'Auinapō," "'auipō," "kulu aumoe," and "pili aumoe" all refer to the late night hours. "Manawa moe" is the time to sleep. The depth of night is known as "pōlalouli," while the gateway connecting midnight and dawn is called "pilipuka." Ibid., 31, 181, 238, 329, 331, 329.
20. Handy, Handy, and Pukui, *Native Planters in Old Hawaii,* 28–29.
21. Kamakau, *Works of the People of Old,* 13.
22. Handy, Handy, and Pukui, *Native Planters in Old Hawaii,* 29; Malo, *Hawaiian Antiquities,* 30.
23. Handy, Handy, and Pukui, *Native Planters in Old Hawaii,* 32.
24. Ibid., 29–30.
25. It is suggested that those who wander about might be born in the month of Nana in the saying "Ua hānau 'ia paha i Nana, ke ma'au ala." Pukui, *'Ōlelo No'eau,* 306.
26. Pukui and Elbert, *Hawaiian Dictionary,* 253.
27. David Turnbull, *Masons, Tricksters and Cartographers: Comparative Studies in the Sociology of Scientific and Indigenous Knowledge* (Amsterdam: Harwood Academic, 2000), 35; Mere Roberts et al., "Whakapapa as a Māori Mental Construct: Some Implications for the Debate over Genetic Modification of Organisms," *Contemporary Pacific: A Journal of Island Affairs* (2004): 1–28; Roberts, "Indigenous Knowledge and Western Science," 65–66; Roma Mere Roberts and Peter R. Wills, "Understanding Maori Epistemology: A Scientific Perspective," in *Tribal Epistemologies: Essays in the Philosophy of Anthropology,* ed. Helmut Wautischer (Burlington, VT: Ashgate, 1998), 43–77.
28. Kame'eleihiwa, *Native Land and Foreign Desires,* 19.
29. Personal communication, Lilikalā Kame'eleihiwa, March 21, 2006.
30. Casey, *Remembering,* 147–149.
31. "Alaloa" also means "to be awake for a long time." This is indicative of ancestral knowledge that continues to live on in Kanaka descendants. Therefore, alaloa conveys both the idea of being on a path and having knowledge that stands the test of time.

Glossary

a'e: indicates an upward or sideways movement; indicates that two things are near or adjacent to one another in space and/or time
'ae kai: place where the waves wash up on the beach
'Aheleakalā: snaring of the sun (*also* 'Aleheakalā and 'Alehelā)
ahiahi: evening
ahu: heap of stones surmounted by an image of a pig; altar carved of kukui wood in the image of a pig
Ahumaunakilo: Observatory Mountain of Ahu
ahupua'a: land division often running from mountain to sea
'ai: to eat; to eat off the land
aia a pa'i 'ia ka maka, ha'i 'ia kupuna nāna 'oe: only when your face is slapped, then you should tell who your ancestors are
aia i ka mole kamali'i, 'a'ohe i o'o ka iwi: still rooted in childhood, the bones have not matured; said of a person who is still childlike
aia i ka 'ōpua ke ola; he ola nui, he ola laulā, he ola hohonu, he ola ki'eki'e: life is in the clouds; great life, broad life, deep life, elevated life
'āina: land; that which feeds
akakū: trance; vision
'ākau: north; right
aku: bonito fish; indicates an action that is going away from the speaker; indicates distant future or past
akua: god, gods
'alae: Hawaiian gallinule or mudhen bird
'Alaehuapī: tricky mudhen
ala hele kūnihi: precarious path

alaloa o Maui: ancestral path around the island of Maui
alanui kīke'eke'e: zigzag road
'Aleheakalā: snaring of the sun (*also* 'Aheleakalā and 'Alehelā)
'Alehelā: snaring of the sun (also 'Aleheakalā and 'Aheleakalā)
ali'i: chief
ali'i 'ai ahupua'a: chief who rules ahupua'a
ali'i 'ai moku: chief who rules a moku
ali'i nī'aupi'o: highest-ranking ali'i
ali'i nui: high-ranking chief
ali'i pi'o: full brother and sister pair born to nī'aupi'o parents; the highest nī'aupi'o mating possible; arching class of ali'i
ali'i wohi: ali'i nui born of a nī'aupi'o parent along with a parent of some other ali'i nui status
aloha 'āina: love for the land
'āluka: scattered places
'ane'i: here
'ano o ka nohona: the nature of one's relationship to one's surroundings or places
ao: enlightenment, to be enlightened; day
'a'ohe e nalo ka iwi o ke ali'i 'ino, 'o ko ke ali'i maika'i ke nalo: the bones of an evil chief will not be hidden, but the bones of a good chief will
'a'ohe kio pōhaku nalo i ke alo pali: on the slope of a cliff, not one jutting rock is hidden from sight
'a'ohe 'oe no ko'u hālau: you are not of my house; a poetic term stating that someone is not related to you
aokanaka: enlightened person
'āpa'a: someone who resides on the same 'āina for an extended period of time
'āpana: land division
'āpapa 'āina: places close to one another
'āpapa mokupuni: places close to one another
au 'āpa'apa'a: ancestral time
'auinalā: afternoon when sun begins to decline
'aumākua: family deities
aumoe: late night, midnight hours
'auwai: ditch
'awa: kava
awakea: time when the sun is high in the sky
'āweoweo: red bigeye fish
'eleao: day
'elepō: night
ēwe: sprout, rootlet, lineage, and kin

hā: breath
hae Hawaiʻi: Hawaiian flag
hahai nō ka ua i ka ulu lāʻau: rains follow the forest
haili moe: dreams
hakaalewa: ladder space
hākilo pono: close observation
haku mele: musical composer
hakuone: small land division
hakupaʻa: small land division; new loʻi
hālau hula: school of hula
Haleakalā: House of the Sun
haliʻa: fond recollection of loved ones
Hāna, mai Koʻolau a Kaupō: Hāna, from Koʻolau to Kaupō; expression defining the boundaries of Hāna
hānai: to adopt, adopted; to feed
hānai hoʻomalu: responsibility of chiefs to feed their people and to provide protection
hānau ka ʻāina, hānau ke aliʻi, hānau ke kanaka: the land is born, the chiefs are born, the general population is born
hāʻueʻue: type of sea urchin
hāʻukeʻuke: type of sea urchin
haumāna: student
he aliʻi ka ʻāina, he kauā ke kanaka: the land is a chief, the people are its servants
he aliʻi no ka malu kukui: a chief of the shade of the kukui; a chief who has something shady in his or her genealogy that he or she doesn't care to discuss
he aliʻi nō mai ka paʻa a ke aliʻi; he kanaka nō mai ka paʻa a ke kanaka: a chief from the class of chiefs; a commoner from the class of commoners
heʻe: octopus
he ʻehu wāwae no ka lani: a trace of the footsteps of a chief; poetic reference to rain and rainbows
he hānai aliʻi, he ʻai ahupuaʻa: one who feeds a chief, rules an ahupuaʻa
hei: string figure
he iʻa moʻa ʻole i kālua: uncooked fish; poetic reference to a low-ranking person
heiau: temple
he kanaka no ka malu kukui: a person from the shade of the kukui tree
he kapu nā pōhaku hānau aliʻi: thunder during a chief's birth is a sign of sanctity
he keiki kāmehaʻi: a wondrous child
he lālā kamahele no ka lāʻau kū i ka pali: a person is of very high rank because of their inaccessibility

he lani i luna, he honua i lalo: a heaven above, an earth beneath

he lupe lele a pulu i ka ua ʻawa: moistened by cold raindrops; a poetic term referring to a person who rises to high ranks

hema: south; left

he maka lehua no kona one hānau: a warrior face of the sands of his/her birth; a term for a person who honored their kulāiwi was compared to a warrior

he māʻona moku: poetic saying to describe aliʻi who were satisfied with their landholdings

he mēheuheu mai nā kūpuna: habits are acquired from one's ancestors

he pali lele a koaʻe: chiefs are like sheer cliffs, not easily approached; poetic reference comparing a high chief to a cliff that is too steep to climb

he pali mania nā aliʻi: the chiefs are sheef cliffs; poetic reference to high chiefs who are not easily accessible

he pili nakekeke: a relationship that rattles

Hīhīmanu: ray fish

hihiʻo: dream

hikianakopili: spittle passed from a dying master

hikina: east

hinupuaʻa: type of banana

hōʻailona: divine omen or sign; royal insignia

hoʻi: more recent term for naha; half brother and half sister pair of nīʻaupiʻo parents

hoʻi hou ka wai i uka o Ao: the water returns again to the upland of Ao

hōʻike a ka pō: visions seen at night

Honuaʻula, e pāluku ʻia ana nā kihi poʻohiwi e nā ʻale o ka Moaʻe: Honuaʻula, whose shoulders are pummeled by the Moaʻe wind; a poetic expression for a person being battered by the wind

Honuaʻula kua laʻolaʻo: callous-backed Honuaʻula

hoʻokele: navigator, navigation

hoʻokohu Kauaʻula, ka makani o ʻUlupaʻu: the Kauaʻula wind of ʻUlupaʻu claims honors that do not belong to it; said in derision of one who steals someone else's genealogy and then brags about illustrious relatives that one is not rightfully entitled to

hoʻopōhaku: figuratively means to live in the same place for generations

hope: back; future; past

hōʻupuʻupu: horrifying visions about someone else

Huʻahuʻakai: sponge; sea foam

hui: joint land ownership; lands held by a group of people whose shares in the parcel are undivided

hula: traditional Kanaka dance
hūpōenui: extremely stupid
iʻa: fish
i ʻāina nō ka ʻāina i ke aliʻi, a i waiwai nō ka ʻāina i ke kanaka: the land is the land because of the chiefs and the land is prosperous because of the general population
i aliʻi nō ke aliʻi i ke kanaka: a chief is a chief because of the people who serve him or her
iʻe kuku: anvil
ʻiewe: placenta
iho: indicates downward motion or an action occurring within oneself, such as thinking, eating, or drinking; indicates near future or recent past
ʻike: to see; to know; to feel, greet, recognize, perceive, experience, be aware, understand; knowledge
ʻike au i kona mau poʻopoʻo: I know all of his or her nooks and crannies; poetic saying referencing someone who is well acquainted with another person
ʻike hānau: knowledge base with which Kānaka are born
ʻikeʻike: knowledge; reduplication of ʻike
ʻike kūhohonu: deep knowledge; insight
ʻike kumu: fundamental knowledge
ʻike kupuna: ancestral knowledge
ʻike loa: to know extensively, to be well versed
ʻike maka: to observe; to identify; to see with one's own eyes
ʻikena: scenery; knowledge
ʻike pāpālua: double knowledge; ability to see the future
ʻike pono: to see and observe well
ʻili: small land division; also known as ʻili ʻāina
ʻili ʻāina: smaller land division than an ahupuaʻa further subdivided into moʻo ʻāina; also known as ʻili
ʻili kūpono: independent land sections; also known as ʻili
inoa ʻāina: place name
inoa pō: name to be given children appearing in dreams
i paʻa i kona kupuna ʻaʻole kākou e puka: had our maternal ancestors died in bearing our grandparents, we would not have been born
ipu: gourd; calabash
i ulu nō ka lālā i ke kumu: the branches grow because of the trunk or source
ʻiwaʻiwa: type of fern
iwi: bone; mountain ridge
kaʻa lalo kūkulu ʻākau: clockwise circuit of an island
kaʻa lalo kūkulu hema: counterclockwise circuit of an island
kahakō: macron; elongated vowel marker

kāhea: to call out

kahikikū: place where the sun rises; east

kahikimoe: place where the sun sets; west; the circle or zone of the earth's surface, whether sea or land, which the eye traverses in looking to the horizon

kahuna: priest, master of an art

kāhuna: priests, masters of arts

kahuna lāʻau lapaʻau: master healer

kai: ocean

Kaihalulu i ke alo o Kaʻuiki: Kaihalulu lies in the presence of Kaʻuiki; poetic saying in reference to the people of Kaihalulu and Kaʻuiki, who are often in the company of one another

kai hī aku: place to cast for bonito fish

kaikaina: younger sibling of the same sex

kai ʻō kilo heʻe: place to spear octopus

kakahiaka: morning

kālaiʻāina: redistribute the land

kālai ʻia: to be carved out

kālai waʻa: canoe builder

kalana: county; smaller division of land than moku; synonymous with okana in some places

Kalani: the heavenly one

ka lani paʻa: the solid heavens

ka lani uli: the solid heavens

ka lewa: region between the lani and ʻāina

kalo: taro

kalo kanu o ka ʻāina: taro planted on the land; a compliment that compared an aliʻi to the kalo of the ʻāina

kama: child

ka Maʻaʻa wehe lau niu o Lele: the Maʻaʻa wind that lifts the coconut leaves of Lele

kamaʻāina: child of the land

kamahele: far-reaching or strong branch

ka makani hāpala lepo o Pāʻia: the dust-smearing wind of Pāʻia

ka makani kā ʻAhaʻaha laʻi o Niua: the gentle ʻAhaʻaha breeze of Niua that drives in the ʻahaʻaha fish

ka makani kokololio o Waikapū: the gusty wind of Waikapū

kamakuʻialewa: region below the lewa lani; the circle of air surrounding the atmosphere

kamaliʻi ʻōkole heleleʻi: loose-seated child

ka malu ao o nā pali kapu o Kakaʻe: the shady cloud of the sacred cliffs of Kakaʻe

ka mokupuni kuapu'u: the hunchbacked island; poetic reference to the island of Maui
kanaka: a native; a person
kānaka: natives; people
Kanaka: Native Hawaiian; used as an adjective for Native Hawaiian(s)
Kānaka: Native Hawaiians
Kanaka 'Ōiwi: Native people of the land
ka nalu he'e o Pu'uhele: the good surf of Pu'uhele
kāne: male; husband or male partner
kanikau: dirge
kanu: to bury, to plant
kanu 'ia: to be planted
ka nui e 'auamo ai i ke keiki i ke kua: the size that enables one to carry a smaller child on the back
ka nui e mo'a ai ka pūlehu: the size that enables one to cook on a fire; poetic reference to a boy who is old enough to have a wife
ka nui e pa'a ai i ka hue wai: the size that enables one to carry a water gourd; poetic reference to a child who is about two years old
ka nui e pa'a ai i nā niu 'elua: the size that enables one to carry two coconuts; poetic reference to a child who is about five years old
kaona: multiple layers of meaning, hidden meaning
ka 'ōpu'u pua i mōhala: a flower that begins to unfold; a baby
kapa: bark cloth
ka pa'a i lalo: the solid below
ka pa'a i luna: the firmament above
ka pae 'āina Hawai'i: the Hawaiian archipelago
ka pae moku: the Hawaiian archipelago
ka pali hinahina o Kā'anapali: the gray hills of Kā'anapali
ka pali kāohi kumu ali'i o 'Īao: the cliff that embraces the chiefly sources of 'Īao
ka pela kapu o Kaka'e: Kaka'e's sacred flesh
Kapōhihihi: branching out of night or chaos
kapu: taboo; sacred
kapu moe: prostrating taboo
kapu noho: crouching taboo
kau: season
ka'u: my, mine
kauā/kauwā: lowest class of ancestral Kanaka society
ka ua Laniha'aha'a o Hāna: rain of the low sky of Hāna
kau ka lā i ka lolo, ho'i ke aka i ke kino: the sun is directly overhead and the shadow of a person retreats back into the body

kaukau ali'i: lesser-ranking chief

kāula: prophet

kaulana mahina: moon calendar; position of the moon

ka 'ulu loa'a 'ole i ka lou 'ia: the breadfruit that even a pole cannot reach; poetic reference to an ali'i nui of very high rank who is like a breadfruit that cannot be reached

ka wai ho'iho'i lā'ī o 'Eleile: the water of 'Eleile that returns back the ti-leaf stalk

ke alanui polohiwa a Kāne: equator; the dark path of Kāne

ke ao ulu: the solid heavens

keapoalewa: the circle of space below Kamaku'ialewa

ke ēwe hānau o ka 'āina: lineal descendants born of the land

keha: height, high, prominent

ke hi'i lā 'oe i ka paukū waena, he neo ke po'o me ka hi'u: when you hold onto the middle section (of a genealogy), the head (beginning) and the tail (ending) of the genealogy are forgotten

kēia: this

keiki: child; children

ke kai holu o Kahului: the swaying sea of Kahului

ke kaulana pa'a 'āina o nā ali'i: the famous landlords of the chiefs; poetic reference meaning the best warriors are awarded the best lands

kēlā: that which is far from the person being spoken to

kēnā: that which is near to the person being spoken to

ki'eki'e: height, tallness

kīhāpai: small land division

kilokilo: to stargaze; stargazer; to read omens; reader of omens

kino lau: physical manifestation of the gods

kīpoho/kīpohopoho: small, shallow pool; small piece of land

kiu: to spy

kō: sugarcane

ko'a: fishing spot, where fish are fed, raised, and harvested

kō'ele: small piece of land

ko kahakai: resources of the ocean

kokoke e 'ā ke ahi o ka 'aulima: almost ready to make fire with a fire stick held in the hand; said of a boy who is almost old enough to mate

ko kula kai: resources of the ocean

ko kula uka: resources of the uplands

komohana: west

komo wai 'ē 'ia: a different liquid had entered; poetic reference to a person whose paternity was suspect

konohiki: chief who managed an ahupua'a; also known as ali'i 'ai ahupua'a

kū: to stand; implies being firmly planted and standing on the ʻāina

kuaʻāina: the backbone of the land; the people who carry the burden of the land on their backs; refers today to people who live off the ʻāina

kuahea: mountain region

kuahiwi: the space below the summit but above the tree line

Kuahiwi Kuauli: dark mountains

kuakua: embankment between loʻi used for cultivation

kua lono: region near the mountaintop; ridge

kua mauna: mountaintop

kūhaka: of high position, as one possessed by the gods

Kūkūau: a type of crab and name of constellation

kukui: candlenut tree

kūkulu: cardinal point

kūkulu ʻehoʻeho: mound of stones

kūkulu o ka lani: border of the sky

kula: plain; source

kulāiwi: plains of ancestral bones; ancestral homeland

kula kai: region near to and including the ocean

kula uka: upland region

kuleana: responsibility; small parcel of land awarded to makaʻāinana in the Māhele, so called because the makaʻāinana had the "responsibility" to care for these lands in perpetuity

kumu: origin, source, foundation; teacher, source of knowledge

Kumuhonua: beginning of the earth

kumu hula: master of the art of hula

Kumulipo: origin of darkness

kumu ʻōlelo Hawaiʻi: Hawaiian language teacher

kūneki nā kūʻauhau liʻiliʻi, noho mai i lalo; hoʻokahi nō, ʻo ko ke aliʻi ke piʻi i ka ʻiʻo: let the lesser genealogies flow away and remain below; let that of the chief ascend

kupa: native

kupa o ka ʻāina: native of the land

kūpapalani: state of heavenly foundation

kupu: to sprout forth and grow

kupua: supernatural being

kupuna: ancestor, elder, grandparent

kūpuna: ancestors, elders, grandparents

kupu wale: spontaneous dream

kuʻu ēwe, kuʻu piko, kuʻu iwi, kuʻu koko: my umbilical cord, my navel, my bones, my blood

lā: sun; day, date; implies distance
lalo: below; surface of the earth; south
Lalohonua: earth below
lalo liloa: region below lalo o ka lepo; abbreviated form of "lalo lilo loa"
lalo lilo loa: region below lalo o ka lepo; also known as "lalo liloa"
lalo o ka lepo: region below lalo
lalo o ka papa kū: region below lalo lilo loa
lani: heavenly one; heaven, sky
lani kua haʻa: poetic name for a very high chief; the highest heaven
lani nuʻu: highest heaven
lau hala: pandanus leaves
Laukīʻeleʻula: dried red ti leaf
lawaiʻa: fisher
lele: type of ʻili ʻāina tied to a larger ahupuaʻa but not physically connected to that ahupuaʻa; to jump
lele kawa: to dive off cliffs
leo: voice
leʻolani: lofty, tall, of chiefly rank
lewa hoʻomakua: the space created when one stands on one foot and raises the other foot
lewa lani: highest stratum of heavens
lewa nuʻu: space in which birds fly
limu: seaweed
lipo: dark night, chaos
līpoa: type of seaweed
loa: long; to be long
loʻi: wetland taro garden
loko: inside
loko iʻa: fishpond
lomilomi: massage
luna: above; north
luna aʻe: the region directly over one's head when standing
luna aku: a region below the luna loa aku
luna lilo aku: a region below the luna lilo loa
luna lilo loa: a region below the luna o ke ao
luna loa aku: a region below the luna lilo aku
luna o ke ao: place where clouds float
Māhele: land division of 1848 between Kamehameha III, the aliʻi, and konohiki that led to the privatization of land in ka pae ʻāina Hawaiʻi
māhele ʻāina: land division

mahina: month
mai: an action toward the speaker; here; this way
mai'a: banana
mai ka hikina a ka lā i Kumukahi a ka welona a ka lā i Lehua: from the rising of the sun in the east at Kumukahi until the setting of the sun in the west at Lehua
mai kaula'i wale i ka iwi o nā kūpuna: do not openly discuss one's mo'okū'auhau and expose the bones of one's kūpuna
mai kekahi kapa a kekahi kapa aku: from one place to another and everything in between
maka'āinana: general population; people who lived off the land
mākāhā: sluice gate
ma ka hana ka 'ike: one learns by actively participating
ma kai: toward the ocean
makaloa: a native sedge
makawalu: scattered places
mākua: parents
makuahine: mother
makua kāne: father
ma lalo iho: just below
mālama 'āina: to care for the land
malo: loincloth
mālua: small land division
ma luna iho: just above
mamao: distant
mamo: descendant
mana: spiritual power
māna: food that has been chewed by a parent for a child
mānaleo: native speaker of 'ōlelo Hawai'i
mā'ona: full, satisfied with enough to eat
mata'einaa: Tahitian word for land divisions similar to ahupua'a
Maui Nui: a large landmass consisting of the islands of Maui, Moloka'i, Lāna'i, and Kaho'olawe, which were all connected in the geologic past
ma uka: toward the mountains
mauliauhonua: descendant of old chiefs of the land
mauna: mountain; the highest places on the landscape
mele: song
mele aloha: love chant
mele ho'ohanohano: honorific chant
mele inoa: name chant
mele ko'ihonua: cosmogonic genealogy

mele wahi pana: song honoring storied places
moana: deep sea
moe ʻawahua: bitter dream
moe hoʻokō: vision that later comes true, fulfilling a prophecy
moe piʻi pololei: prophetic dream
moe ʻuhane: dream
mōʻī: monarch
moku: large land division or district
mokuʻāina: district; an island
moku kele i ka waʻa: an island that needed to be sailed to
mokulua: two adjacent islets
mokuoloko: large land division, district
mokupuni: island; places that were cut off from other lands by the sea
mole: taproot; foundation
moʻo: lizard, reptile, dragon; connection; succession, series, especially a genealogical line, lineage, successive line; narrow strips of farmed land; a demigod found near water sources, often compared to mermaids in a modern context
moʻo ʻāina: land division that was subdivided into paukū ʻāina
moʻo aliʻi: genealogy of aliʻi
moʻo hele: path
moʻo helu: enumerate; recite
moʻo kaʻao: story
moʻo kanaka: succession of people
moʻokūʻauhau: genealogy; genealogical accounts
moʻo kupuna: ancestral genealogy
moʻo lau: many descendants
moʻolelo: historical account; history, literature, narrative
moʻopuna: grandchildren; descendants
mua: front; future; past
naʻau: intestines, bowels, guts; mind, heart, affections of the heart or mind; mood, temper, feelings
naʻau ahonui: patient
naʻau aliʻi: kind, thoughtful
naʻau aloha: benevolent
naʻauao: enlightened person
naʻauʻauā: intense grief
naʻau hoʻokiʻekiʻe: conceited
naʻau ʻinoʻino: nasty-hearted
naʻau keʻemoa: evil-hearted

naʻau kopekope: spiteful

naʻau kūhili: careless, thoughtless

naʻau lua: indecisive

naʻau palupalu: soft-hearted

naʻau pēpē: unpretentious

naʻaupō: one whose intestines are dark; to be ignorant

naʻau pono: just; upright

naha: half brother and half sister pair of nīʻaupiʻo parents; curved

Nāhiku hauwalaʻau: Nāhiku of the loud-talking people

Nā Hono a Piʻilani: the Bays of Piʻilani, a chief of Maui

na kahi ka malo, na kahi e hume: the loincloth belongs to one, another wears it; poetic reference to a very close blood relationship

Nā Kai ʻEwalu: the eight seas, poetic reference to the Hawaiian Islands; eight channels as seen from Lahaina, Maui; eight cardinal points

nā keiki uneune māmane o Kula: the children of Kula who tug and pull up the māmane; poetic reference to the diligent people of Kula who are praised for accomplishing their goals

nā lālā kapu a Lono: the sacred branches or descendants of Lono

nā mea e hoʻopuni ana: everything that surrounds or encircles a person

nanaʻe: small piece of land

nānā ka maka; hana ka lima: observe with the eyes; work with the hands

nānā ka maka; hoʻolohe ka pepeiao; paʻa ka waha: observe with the eyes; listen with the ears; shut the mouth

nā pela kapu o Kakaʻe: Kakaʻe's sacred flesh

Nā Poko: a moku on Maui meaning "the smaller land divisions"

Nā Wai ʻEhā: the four waters; poetic reference to Waikapū, Wailuku, Waiʻehu, and Waiheʻe

na wai hoʻi ka ʻole o ke akamai, he alanui i maʻa i ka hele ʻia e oʻu mau mākua: who does not have the intelligence to travel a path that has been well traversed by one's ancestors?

nei: indicates close proximity

nīʻaupiʻo: arching of the coconut frond; highest-ranking aliʻi class

niu: coconut

noiʻi i nā mea ʻike ʻole ʻia: seeker of unseen things

no ka noho ʻāina ka ʻāina: the land is for those who reside on the land

noʻonoʻo mua: dream about previous experiences

ʻō: there

ʻohana: family

ʻohe: bamboo

ʻōiwi: native
ʻo ka ʻaui aku nō koe o ka lā: the sun will soon go down; poetic reference to the final season of one's life
ʻokana: sections that had been cut off; sometimes defined as a smaller land division than a kalana or a mokuʻāina
ʻo ke aliʻi ka mea ikaika, ʻaʻole ʻo ke kanaka: it is the chief who is strong, not the general population
ʻokina: glottal stop
ola ka inoa: the name lives
ʻōlelo: language
ʻōlelo Hawaiʻi: Hawaiian language
ʻōlelo Kahiki: Tahitian language
ʻōlelo makuahine: mother tongue of Hawaiʻi
ʻōlelo noʻeau: wise saying, proverb
oli: chant
olōlo: brains or oily coconut meat
one hānau: sands of birth
one kōʻele: small piece of land
ʻopihi: limpet
paʻa: to be steadfast and committed to one's place
paʻa ka waha, hana ka lima: close your mouth, work with your hands
pae ʻāina: Hawaiian archipelago
pāʻeli: to dig the earth, as in planting in a loʻi
pae moku: Hawaiian archipelago
pākahi ka nehu a Kapiʻioho: the nehu of Kapiʻioho are rationed, one to a person
Palikū: vertical precipice
Papa: earth mother
Papahānaumoku: Papa that gives birth to islands
paukū ʻāina: small land division
piko: umbilical cord
pili: type of grass
piʻo: arching classes of aliʻi; high-ranking aliʻi
pō: night
Pōʻalima: Friday
pōhaku: stone
poi: mashed, cooked taro thinned with water
poʻina kai: place where waves break
pō mahina: night
pono: virtuous; good; harmony; balance

poʻolua: double paternity

pua ka neneleau, momona ka wana: the neneleau blooms and the sea urchin is fat; poetic reference to the time when sea urchins are ready to be harvested

pua ke kō, kū ka heʻe: the sugarcane tassels bloom, the octopus appears; late October, early November

Pūʻali Komohana: west of the Maui isthmus

Puanue: rainbow

pueo: owl

pulapula: offspring, descendant

pūloʻuloʻu: insignia of chiefly kapu

puluwai: small piece of land

puna: spring; poetic for source of knowledge

pūnāwai: spring; metaphoric reference to spring of knowledge

pūnāwai waiwai: rich repository of wisdom

ua hānau ʻia paha i Nana, ke maʻau ala: wanderers are born in the month of Nana

ua hele au i kēia mau kuahiwi a lewa: I have traveled these mountains so extensively that I know every nook and cranny

ʻuala: sweet potato

ua noho au a kupa: I have resided until well acquainted with this place

ua puka a maka: face emerges and is seen

uka: mountain region; upland

ʻūlāleo: unusual supernatural sound

ulua: type of fish, giant kingfish or giant trevally

wā: period of time

waena: middle

waho: outside

wai: water

Waiʻehu mai ka pali o Kapulehua a ka pali o ʻAʻalaloa: Waiʻehu extending from the cliff of Kapulehua to the cliff of ʻAʻalaloa

Wailua: spirit

waiwai: wealth

Wākea: sky father

wanaʻao: dawn

wao akua: upland region reserved for the gods

wao ʻeiwa: an inland region

wao kanaka: an inland region that people frequent

wao lipo: an inland region

wao maʻukele: an upland region

wao nahele: an inland region; a forested region

wauke: paper mulberry

Welehu ka malama, liko ka ʻōhiʻa: Welehu is the month when the ʻōhiʻa trees are putting forth leaf buds

Bibliography

Abbott, Isabella Aiona. *Lā'au Hawai'i: Traditional Hawaiian Uses of Plants*. Honolulu: Bishop Museum Press, 1992.

Alexander, W. D. *Appendix 1, to Report: A Brief History of Land Titles in the Hawaiian Kingdom*. Honolulu: P. C. Advertsier Co., 1882.

———. *A Brief History of the Hawaiian People*. New York: American Book Co., 1920.

Andrade, Carlos. *Hā'ena: Through the Eyes of the Ancestors*. Honolulu: University of Hawai'i Press, 2008.

Andrews, Lorrin. *A Dictionary of the Hawaiian Language*. Waipahu: Island Heritage Publishing, 2003.

Armstrong, Nanea K. "Moving Time and Land: Cultural Memories through Inscribed Landscapes." M.A. thesis, University of Hawai'i at Mānoa, 2003.

Barrère, Dorothy B. "Cosmogonic Genealogies of Hawaii." *Journal of the Polynesian Society* 70:4 (1961): 419–428.

Beamer, Nona. *Nā Mele Hula*. Vol. 2: *Hawaiian Hula Rituals and Chants*. Lā'ie: Institute for Polynesian Studies, 2001.

Beckwith, Martha Warren. *Hawaiian Mythology*. Honolulu: University of Hawai'i Press, 1970.

———. *The Kumulipo: A Hawaiian Creation Chant*. Honolulu: University of Hawai'i Press, 1972.

Bishop Museum Archives. HEN. Vol. 1, Hms K9 Kalaniana'ole Collection.

———. HI.L1.3#1.

———. HI.L1.3#4.

———. HI.L1.3#5.

———. HI.L8.

———. MS SC Emory Group 8, Box 4.7.

———. MS SC Handy, Esc, Box 7.30.

Buke Mahele.

Cajete, Gregory. *Native Science: Natural Laws of Interdependence*. Santa Fe, NM: Clear Light Publishers, 2000.

Casey, Edward S. "How to Get from Space to Place in a Fairly Short Stretch of Time: Phenomenological Prolegomena." In *Senses of Place*, ed. Steven Feld and Keith H. Basso, 13–52. Santa Fe, NM: School of American Research Press, 1996.

———. *Remembering: A Phenomenological Study*. Bloomington: Indiana University Press, 2000.

Culliney, John L. *Islands in a Far Sea: The Fate of Nature in Hawai'i*. Honolulu: University of Hawai'i Press, 2006.

Davis, T., T. O'Regan, and J. Wilson. *Nga Tohu Pumahara: The Survey Pegs of the Past*. Wellington: New Zealand Geographic Board, 1990.

Elbert, Samuel H., and Mary Kawena Pukui. *Hawaiian Grammar*. Honolulu: University of Hawai'i Press, 1979.

Fornander, Abraham. *An Account of the Polynesian Race: Its Origins and Migrations and the Ancient History of the Hawaiian People to the Times of Kamehameha I*. Whitefish, MT: Kessinger Publishing, 2007.

———. *Fornander Collection of Hawaiian Antiquities and Folk-Lore: The Hawaiian Account of the Formation of Their Islands and Origin of Their Race, with the Traditions of Their Migrations, Etc., as Gathered from Original Sources*. Ed. Thomas G. Thrum. Facsimile ed. Vols. 4 and 6. Honolulu: 'Ai Pōhaku Press, 1999.

Handy, E. S. Craighill, Elizabeth Green Handy, and Mary Kawena Pukui. *Native Planters in Old Hawaii: Their Life, Lore, and Environment*. Honolulu: Bishop Museum Press, 1972.

Hau'ofa, E. "Our Sea of Islands." *Contemporary Pacific: A Journal of Island Affairs* 6:1 (1994): 148–161.

Hawaiian Historical Society. *Annual Report of the Hawaiian Historical Society*. Honolulu: Hawaiian Historical Society, 1904.

"He Kanikau no ko Maua Papa Heleloa." *Ka Nupepa Kuokoa*. August 21, 1924.

"He Mau Mea i Hoohalahala ia no na Mea iloko o na Kaao Hawaii." *Ka Nupepa Kuokoa*. February 15, 1868.

Hill, Deborah. "Distinguishing the Notion 'Place' in Oceanic Language." In *The Construal of Space in Language and Thought*, Martin Pütz and René Dirven, eds., 307–328. Berlin: Mouton de Gruyter, 1996.

Hiroa, Te Rangi (Sir Peter H. Buck). *Arts and Crafts of Hawaii*. Honolulu: Bishop Museum Press, 2003.

Hospice Hawai'i. *Volunteer Handbook: Resources, Procedures and Guidelines*. Honolulu: Hospice Volunteer Services, 2011.

Ii, John Papa. *Fragments of Hawaiian History*. Trans. Mary K. Pukui. Ed. Dorothy B. Barrère. Rev. ed. Honolulu: Bishop Museum Press, 1983.

Johnson, Jay T., Renee Pualani Louis, and Albertus Hadi Pramono. "Facing the Future: Encouraging Critical Cartographic Literacies in Indigenous Communities." *ACME: An International E-Journal for Critical Geographies* 4:1 (2006): 80–98.

Johnson, Rubellite. *Essays in Hawaiian Literature.* Part 1: *Origin Myths and Migration Traditions.* Honolulu: University of Hawai'i Press, 2001.

———. Hawaiian 261. (unpublished manuscript). Honolulu: University of Hawai'i at Mānoa, 2001.

———. *Kumulipo: Hawaiian Hymn of Creation.* Vol. 1. Honolulu: Topgallant Publishing, 1981.

Jones, Stephen B. "Geography and Politics in the Hawaiian Islands." *Geographical Review* 28:2 (April 1938): 193–213.

Judd, Henry P. *The Hawaiian Language and Hawaiian-English Dictionary.* Honolulu: Mutual Publishing, 1995.

Kamakau, Samuel Manaiakalani. "Ka Moolelo Hawaii." *Ke Au Okoa,* May 13, 1869, to December 22, 1870.

———."Ka Moolelo o Kamehameha I." *Ka Nupepa Kuokoa,* October 20, 1866, to March 30, 1867.

———. *Ka Po'e Kahiko: The People of Old.* Trans. Mary Kawena Pukui. Ed. Dorothy B. Barrère. Bernice P. Bishop Museum Special Publication No. 51. Honolulu: Bishop Museum Press, 1991.

———. *Ruling Chiefs of Hawaii.* Rev. ed. Honolulu: Kamehameha Schools Press, 1992.

———. *Tales and Traditions of the People of Old: Nā Mo'olelo O Ka Po'e Kahiko.* Honolulu: Bishop Museum Press, 1993.

———. *The Works of the People of Old: Nā Hana a Ka Po'e Kahiko.* Translated by Mary K. Pukui. Edited by Dorothy B. Barrère. Honolulu: Bishop Museum Press, 1976.

Kame'eleihiwa, Lilikalā. *Native Land and Foreign Desires: How Shall We Live in Harmony: Ko Hawai'i 'Āina a Me Nā Koi Pu'umake a Ka Po'e Haole: Pehea Lā E Pono Ai.* Honolulu: Bishop Museum Press, 1992.

"Ka Mookuauhau Alii a WM. U. Hoapili Kanehoa." *Ka Nupepa Kuokoa,* August 15, 1913.

"Ka Moolelo o Hawaii Nei." *Ka Nupepa Kuokoa,* July 29, 1865, August 5, 1865, and September 30, 1865.

Ka Ohu Haaheo i na Kuahiwi Ekolu. "Ka Waiu Ame Ka Meli." *Ka Hoku o Hawaii,* June 17, 1920.

Kanahele, George S. *Kū Kanaka, Stand Tall: A Search for Hawaiian Values.* Honolulu: University of Hawai'i Press, 1986.

Kanahele, Ku'ulei Higashi, ed. *Mauiloa: A Collection of Mele for the Islands of Maui, Kaho'olawe, Lāna'i, Moloka'i.* Trans. Pualani Kanaka'ole Kanahele. Honolulu: Alu Like, 2005.

Kanepuu, J. H. "Kaahele ma Molokai: Helu 5." *Ke Au Okoa,* October 17, 1867.

Kaulia, Jas. K. “He Adamu No Iloko O Ka Lahui Hawaii.” *Ke Aloha Aina*, June 6, 1896.

Keaulana, Kimo ‘Alama. “Koali.” In Mele Collection of Kimo Alama Keaulana, Bishop Museum Archives.

Kelly, Jan. “Maori Maps.” *Cartographica* 36:2 (1999): 1–30.

Kepelino and Martha Warren Beckwith. *Kepelino's Traditions of Hawaii*. Bernice P. Bishop Museum Bulletin 95 (1932). Honolulu: Krause Reprint, 1971.

King, R. D. “Districts in the Hawaiian Islands.” In J. W. Coulter, comp., *A Gazetteer of the Territory of Hawaii*. University of Hawai‘i Research Pub. No. 11, 1935.

Landgraf, *Nā Wahi Kapu o Maui*. Honolulu: ‘Ai Pōhaku, 2003.

Liliuokalani. *The Kumulipo: An Hawaiian Creation Myth*. Kentfield, CA: Pueo Press, 1997.

Lucas, Paul F. Nahoa. *A Dictionary of Hawaiian Legal Land-Terms*. Honolulu: Native Hawaiian Legal Corporation and University of Hawai‘i Committee for the Preservation and Study of Hawaiian Language, Art, and Culture, 1995.

Lyons, C. J. “Land Matters in Hawaii.” *The Islander,* issues from July 9 to July 30, 1875.

MacKenzie, Melody Kapilialoha. *Native Hawaiian Rights Handbook*. Honolulu: Native Hawaiian Legal Corporation, 1991.

Malo, David. *Hawaiian Antiquities: Mo‘olelo Hawai‘i*. Honolulu: Folk Press, 1987.

Maunupau, Thomas K. *Huakai Makaikai a Kaupo, Maui: A Visit to Kaupō, Maui: As Published in Ka Nupepa Kuokoa, June 1, 1922–March 15, 1923*. Honolulu: Bishop Museum Press, 1998.

McKinzie, Edith Kawelohea. *Hawaiian Genealogies: Extracted from Hawaiian Language Newspapers*. Vol. 2. Honolulu: University of Hawai‘i Press, 2003.

Mokuleia, S. “Kekahi Mau Mea a ka Poe Kahiko i Mahele ai o Hawaii Nei i ka Wa Kahiko.” *Ka Nupepa Kuokoa*, March 7, 1868.

“Mookuauhau Alii: Na Iwikuamoo o Hawaii Nei Mai Kahiko Mai.” *Ka Makaainana,* April 27, 1896, to November 16, 1896.

“Moolelo Hawaii.” *Ka Hoku o Hawaii,* December 21, 1911.

Nakuina, Moses K. *Moolelo Hawaii O Pakaa a Me Ku-a-Pakaa*. Honolulu (privately printed), 1902.

“Na Lei o Hawaii.” *Ka Nupepa Kuokoa,* February 24, 1883.

“Na Papa Alii o Hawaii.” *Ka Nupepa Kuokoa,* March 10, 1883.

“No Ka Wehewehe Ana.” *Ka Hae Hawaii,* March 12, 1856.

Peleioholani, S. L. “The Ancient History of Hookumu-Ka-Lani = Hookumu-Ka-Honua.” Manuscript in Bishop Museum Archives, HI.L.1.3.1.

Poepoe, J. M. “Ka Moolelo Hawaii Kahiko: Mokuna I Na Kuauhau Kahiko e Hoike ana i na Kumu i Loaa ai ka Pae Moku o Hawaii Nei.” *Ka Na'i Aupuni,* February 3 to August 4, 1906.

Preza, Donovan. "The Empirical Writes Back: Re-examining Hawaiian Dispossession Resulting from the Mahele of 1848." M.A. thesis, University of Hawai'i at Mānoa, 2010.

Puaaloa. "Ka Moolelo O Maui." *Ka Nupepa Kuokoa*, June 27 and July 4, 1863.

Pukui, Mary Kawena. *'Ōlelo No'eau: Hawaiian Proverbs and Poetical Sayings.* Honolulu: Bishop Museum Press, 1983.

Pukui, Mary Kawena, and Samuel H. Elbert. *Hawaiian Dictionary: Hawaiian-English, English-Hawaiian.* Rev. and enl. ed. Honolulu: University of Hawai'i Press, 1986.

Pukui, Mary Kawena, Samuel H. Elbert, and Esther T. Mookini. *Place Names of Hawai'i.* Rev. and enl. ed. Honolulu: University Hawai'i Press, 1976.

Pukui, Mary Kawena, E. W. Haertig, and Catherine A. Lee. *Nānā I Ke Kumu (Look to the Source).* Two vols. Honolulu: Hui Hanai, 1983.

Roberts, Mere. "Indigenous Knowledge and Western Science: Perspectives from the Pacific." *Royal Society of New Zealand: Miscellaneous Series* 50 (1996): 69–75.

Roberts, Mere, Brad Haami, Richard Benton, Terre Satterfield, Melissa L. Finucane, Mark Henare, and Manuka Henare. "Whakapapa as a Māori Mental Construct: Some Implications for the Debate over Genetic Modification of Organisms." *Contemporary Pacific: A Journal of Island Affairs* 16:1 (2004): 1–28.

Roberts, Roma Mere, and Peter R. Wills. "Understanding Maori Epistemology: A Scientific Perspective." In *Tribal Epistemologies: Essays in the Philosophy of Anthropology*, ed. Helmut Wautischer, 43–71. Burlington, VT: Ashgate, 1998.

Schattenburg-Raymond, Lisa. "Nā Waiho'olu'u O Ke Ānuenue." Song in *Ka 'Aha Hula 'O Hālauaola.* Honolulu: Maui Arts and Culture Center and Maui Community College, 2005.

Silva, Noenoe K. *Aloha Betrayed.* Durham and London: Duke University Press, 2004.

State of Hawai'i Division of Forestry and Wildlife and State of Hawai'i Department of Land and Natural Resources. *Wao Akua: Sacred Source of Life.* Honolulu: Department of Land and Natural Resources, 2003.

Sterling, Elspeth P., and Native Hawaiian Culture and Arts Program. *Sites of Maui.* Honolulu: Bishop Museum Press, 1998.

Tuan, Yi-fu. "Language and the Making of Place: A Narrative-Descriptive Approach." *Annals of the Association of American Geographers* 81:4 (1991): 684–696.

Turnbull, David. *Masons, Tricksters and Cartographers: Comparative Studies in the Sociology of Scientific and Indigenous Knowledge.* Amsterdam: Harwood Academic, 2000.

Waiamau, J. "Ka Hoomana Kahiko Helu 27: Na Papa O Kanaka, Na'lii, Na Makaainana, Na Lopa, Na Hu, Na Kauwa." *Ka Nupepa Kuokoa,* November 11, 1865.

Westervelt, W. D. *Hawaiian Historical Legends.* New York: Fleming H. Revell Co., 1923.

———. *Legends of Maui—A Demi-God of Polynesia and of His Mother Hina.* Melbourne: G. Robertson & Co., 1913.

Wilcox, Carol, Kimo Hussey, Vicky Hollinger, and Puakea Nogelmeier. *He Mele Aloha: A Hawaiian Songbook*. Honolulu: ʻOliʻoli Productions, 2004.
Wise, John H. "He Moolelo No Ka Hookumuia Ana O Na Paemoku O Hawaii Nei a Me Ka Hoolaukanaka Ana I Hoikeia Ma Na Mele Hawaii Kahiko." *Ke Au Hou,* January 24 and February 14, 1912, 3 vols.
———. "The History of Land Ownership in Hawaii." Ed. Kamehameha Schools. Rutland, VT, & Tokyo, Japan: Charles E. Tuttle, 1965.
Woodward, David, and G. Malcolm Lewis. *The History of Cartography: Cartography in the Traditional African, American, Arctic, Australian, and Pacific Societies*. Vol. 2. Chicago: University of Chicago Press, 1998.

Index